INTRODUCTION TO SENSOR SYSTEMS

The Artech House Communication and Electronic Defense Library

Principles of Secure Communication Systems by Don J. Torrieri
Introduction to Electronic Warfare by D. Curtis Schleher
Electronic Intelligence: The Analysis of Radar Signals by Richard G. Wiley
Electronic Intelligence: The Interception of Radar Signals by Richard G. Wiley
Pulse Train Analysis Using Personal Computers by Richard G. Wiley and Michael B. Szymanski
RGCALC: Radar Range Detection Software and User's Manual by John E. Fielding and Gary D. Reynolds
Signal Theory and Random Processes by Harry Urkowitz
Signal Theory and Processing by Frederic de Coulon
Digital Signal Processing by Murat Kunt
Analysis and Synthesis of Logic Systems by Olis Rubin
Advanced Mathematics for Practicing Engineers by Kurt Arbenz and Alfred Wohlhauser
Mathematical Methods of Information Transmission by Kurt Arbenz and Jean-Claude Martin
Codes for Error-Control and Synchronization by Djimitri Wiggert
Machine Cryptography and Modern Cryptanalysis by Cipher A. Deavours and Louis Kruh
Cryptology Yesterday, Today, and Tomorrow, Cipher A. Deavours et al., eds.
Microcomputer Tools for Communications Engineering by S.T. Li, J.C. Logan, J.W. Rockway, and D.W.S. Tam

INTRODUCTION TO SENSOR SYSTEMS

S.A. HOVANESSIAN

ARTECH HOUSE

Library of Congress Cataloging-in-Publication Data

Hovanessian, Shahan A., 1931 —
Introduction to sensor systems.

Bibliography: p.
Includes index.
1. Radar. I. Title.
TK6576.H6728 1988 621.3848 88-22135
ISBN 0-89006-271-4

ARTECH HOUSE, INC.
685 Canton Street
Norwood, MA 02062

International Standard Book Number: 0-89006-271-4
Library of Congress Catalog Card Number: 88-22135

10 9 8 7 6 5 4 3 2 1

Contents

Preface

This textbook contains introductory material on several types of sensor systems currently in operation for surveillance, ground mapping, and tracking and recognition of targets. Specifically, the contents of the book include a discussion of microwave radar sensors, millimeter wave radar systems, electro-optical thermal imagers, infrared search and tracking systems, and laser radar systems. Additionally, multiple sensor system data fusion and tracking, as well as evaluation of the effectiveness of these systems are discussed. The book is intended for engineers and scientists who desire a working knowledge of the design parameters, components, and performance characteristics of these sensors. The operational environment and limitations in detection range and angular position accuracy of each sensor system is discussed as well.

Generally, sensor systems are divided into two broad categories: active and passive sensors. Active sensors are the ones that radiate their own energy for target detection, an example being the microwave radar. Passive systems, on the other hand, rely on natural or human-made radiation from the target for detection. Thermal imagers, for example, use the natural radiated heat energy of the target and its thermal differences from background to produce images. Human-made sources of radiation, such as the radiated heat from exhaust discharge of engines, are also used for target detection, as in the case of infrared search and tracking systems.

Another important parameter separating the categories of sensors is their operating frequency (wavelength). Microwave radars operate in the gigahertz (10^9 Hz) frequency region that corresponds to centimeter wavelengths. Millimeter-wave radars operate in the tens to hundreds of gigahertz corresponding to wavelengths, as the name implies, in the millimeter region. The electro-optical systems, which include thermal imagers, infrared search and tracking systems, and laser radars, operate in the micron (10^{-6} m) wavelengths region. Atmospheric attenuation effects, a strong function of the operating wavelength, affect the performance of various sensor systems differently, as discussed in detail in this book.

The chapters of the book are arranged in the following order. Chapter 1 gives a complete overview of various sensor systems, their operating conditions and limitations, and differences in performance among sensors. Chapter 2 gives a discussion of microwave radar systems and principles involved in the design and operation of these systems. Chapter 3 describes the very interesting microwave radar application of radar imaging and mapping. We show that by using radar principles, large areas of the ground can be imaged from airborne and spaceborne radars. Chapter 4 gives a discussion of millimeter-wave radars. Although they operate at higher frequencies than do microwave radars, most of the radar principles covered in the previous chapters apply to these radars. The atmospheric effects, however, are different from those of microwave radars, primarily because of the wavelengths involved.

Chapter 5 gives a discussion of thermal imaging systems and their design and performance characteristics. These imaging systems operate on the temperature gradient of the object scene and produce a television-like picture of the mapped area. Because they operate on temperature gradient, as opposed to light reflection in photographic systems, they can produce images at night. For this reason they are also termed *night vision systems*. This discussion begins with electro-optical systems and expands to include infrared search and tracking systems. These systems also use heat sensing detectors, but they are generally designed for large volume search and detection of targets ("hot spots") against the sky or a clear background. These targets may be flying aircraft, missiles, and so on. Chapter 6 gives the performance equations of laser radar systems. These systems, like the thermal imagers, operate in the micron wavelength region of the spectrum. They are, however, active devices similar to microwave radars, as they use their own source of radiated energy for target detection.

The last two chapters of the book are devoted to multiple sensor systems. Chapter 7 gives a discussion of multiple sensor tracking and data fusion. This is a new field as, only recently, combinations of sensors are being used to improve target detection and tracking capabilities. In Chapter 8, methods of quantitative multiple-sensor effectiveness evaluation are discussed. In particular, a new method, titled the *expert matrix method,* is introduced for the quantitative effectiveness evaluation of multiple sensors against a given set of requirements. The expert matrix method of this chapter can also be applied for quantitative evaluation of options among several proposals, designs, and systems.

S. A. HOVANESSIAN
LOS ANGELES
JULY 4, 1988

ACKNOWLEDGEMENTS

During the preparation of this book I had the benefit of numerous discussions with a number of colleagues who were specialists in the specific sensors covered in this book. These discussions served to improve the level of the contents of the book as well as simplify the explanation of the technical principles involved. In particular, my appreciation is extended to Dr. David L. Glackin for his comments on the thermal imaging and infrared chapters, to Dr. Robert B. Dybdal for his review of the millimeter wave radar chapter, to Dr. Nemesio Caraballo for his review of the laser radar chapter and to Mr. J. S. Avrin for his general comments on the inclusion of examples of various sensor systems.

Multiple target tracking and sensor fusion, Chapter 7, was written by Mr. Samuel S. Blackman. This field of study is relatively new and applicable concepts are not as yet formalized. Considering this fact, the efforts of Mr. Blackman in combining the present knowledge on the subject and writing this chapter in a cohesive manner are commendable.

The general support of the Aerospace Corporation and, in particular, Mr. J. R. Parsons, Principal Director, Sensor Systems Subdivision, and Mr. Robert S. Gaylord, General Manager, Electronics and Optics Division, were of great value in the preparation of this textbook.

The descriptive pictures of sensor systems shown in this textbook were obtained primarily from manufacturers' brochures and open trade publications. The author extends his appreciation to several of these companies including: Hughes Aircraft Company, Datron Corporation, Raytheon Company, Ferranti Defense Systems, DFVLR Federal Republic of Germany, and MacDonald Dettwiler Corporation.

Chapter 1
Introduction to Sensor Systems

This chapter introduces several categories of sensors that are used for detection, tracking, recognition, and identification of targets. These sensors are compared as to their mode of operation, operating frequency, and resolution considerations. The design, performance, and application of each sensor system is briefly discussed in this chapter. The detailed description of individual sensors is covered in future chapters. Sensors covered specifically include microwave radars, millimeter-wave radars, electro-optical thermal imagers, and laser radar systems. Note that the electro-optical systems discussed in this book consist of thermal imagers, infrared search and tracking systems, and laser radars. Visual sensor systems of TV-like and ultraviolet sensors are not discussed.

1.1 INTRODUCTION

Sensor systems are constructed for a variety of applications that include surveillance and tracking of targets, instrumentation, air traffic control, ground mapping and imaging, and detection and recognition of objects. Some of these sensors provide their own source of target illumination and are referred to as *active* sensors. Microwave and laser radars are in this category. *Passive* sensors do not provide target illumination. They depend on natural conditions for detection. A common example of this is photographic cameras that use reflected light energy as their source of "target" illumination. Infrared detectors are also passive devices that use radiated heat energy of the natural background and human-made objects as their source of detection energy. For this reason they can operate at night as well as in the day time.

Active systems are equipped with both a transmitting and a receiving system. Passive systems, on the other hand, consist of only a receiving system. In a great number of active systems, the transmitting and receiving systems share the antenna (or the lens) between transmitting and receiving functions.

In this chapter we introduce several sensor systems as to their operational modes and their performance capabilities. Each of these sensors will be discussed in greater detail, with supporting performance equations and numerical examples, in future chapters. The sensor systems considered are microwave radars, *millimeter-wave* (MMW) radars, and *electro-optical* (EO) systems. The EO systems also include infrared systems, thermal imagers, and laser radar systems. These sensor categories are primarily separated by applicable wavelengths.

Table 1.1 compares microwave, MMW, and EO sensor systems. From this table, note that the microwave systems operate in the centimeter wavelength region, MMW radars operate in the millimeter wavelength, and the EO systems utilize micrometer wavelength; that is, 10^{-6} m.

Table 1.1 Comparison of Microwave, MMW, and EO Sensors

Condition	*Microwave*	*MMW*	*EO*
Wavelength	cm (10^{-2} m)	mm (10^{-3} m)	μm (10^{-6} m)
Components	Large	Small	Small
Aperture (angular resolution)	Large (degrees)	Small (mr)	Small (μr)
Mode	Active	Active	Passive-active
Weather conditions, dust and smoke	No degradation	Some degradation	Degraded

The size of components used in these sensor systems—transmitters, waveguides, detectors, receivers, and so on—are usually a function of the wavelength. Microwave systems, because of their larger wavelengths, have larger components. The components of MMW radars and EO systems are considerably smaller.

An important part of any sensor system is its aperture, where transmitted (received) energy leaves (enters) the system. Microwave and MMW radar apertures are their antennas, while EO system apertures consist of optical lenses and mirrors.

The gain of antennas and lenses, which is a measure of directivity of the transmitted (received) energy, can be written as

$$G = \frac{4\pi A}{\lambda^2} \tag{1.1}$$

where G is the gain, A is the aperture area, and λ is the wavelength. From this equation, for a given aperture area, smaller wavelengths will produce large gains. Conversely, larger wavelengths will require larger apertures to produce a given gain. As shown in Table 1.1, microwave systems usually have large apertures as compared to MMW and EO systems.

As mentioned earlier, the gain of an antenna is a measure of its ability to direct the transmitted energy in a given direction. This is better illustrated by writing the gain equation in the following form:

$$G = \frac{4\pi}{\Theta_{BW}^2} \tag{1.2}$$

where 4π is the area of a sphere of unit radius in steradians and Θ_{BW} is the antenna beamwidth, or the lens diffraction limit, in units of radians.

Thus, in the case of a microwave antenna with a 2° beamwidth, (1.2) results in a gain of

$$\begin{aligned} G &= \frac{4\pi}{(2/57.3)^2} \\ &= 10314 \\ &= 40.13 \text{ dB} \end{aligned} \tag{1.3}$$

where 57.3 is the conversion from degrees to radians.

The value of Θ_{BW} can also be written in terms of wavelength λ and antenna dimension D as follows:

$$\Theta_{BW} = \frac{\lambda}{D} \tag{1.4}$$

From (1.4), a CO_2 laser radar with a λ of 10.6 μm and lens diameter of 20 cm will result in a beamwidth of

$$\begin{aligned} \Theta_{BW} &= \frac{10.6 \times 10^{-6}}{20 \times 10^{-2}} \\ &= 5.3E - 5 \text{ rad} \\ &= 53\ \mu\text{r} \end{aligned} \tag{1.5}$$

Note that the gain and beamwidth equations (1.2) and (1.4), respectively, are approximations to these values. Actual gain and beamwidth values may vary by as much as 30 percent from these approximations depending on the physical structure and electromagnetic properties of antennas and lenses. From (1.4), note that larger wavelengths, for a fixed size aperture, will result in wider beam dispersions. This beam dispersion Θ_{BW} is directly related to the system angular resolution as shown in Table 1.1. From this table, using nominal parameters, microwave radars will have beamwidths in degrees, MMW radars in milliradians, and EO systems in microradians, as shown.

The operating mode of microwave and MMW radars is active, as these sensor systems are equipped with transmit and receive systems. Electro-optical thermal imaging and infrared search and track systems are passive devices. Laser radars, operating in the same EO frequency band, are active sensing devices.

The last item of Table 1.1 consists of sensor operation under degraded conditions. Degraded conditions may be due to weather, such as rain or fog, or road conditions of dust and smoke. In general, when the wavelengths are in the same order of magnitude as the weather particles of fog and rain or the dust and smoke, the atmospheric attenuation of the target radiated energy will be considerable. For this reason, microwave radars with wavelengths in centimeters are not degraded under adverse atmospheric conditions. Electro-optical systems with micrometer wavelengths are degraded under adverse weather conditions. The MMW radars are also affected by adverse weather but not to the extent of EO systems.

1.2. SENSOR SYSTEM FREQUENCIES

The frequencies of different sensor systems are shown in Figure 1.1, together with corresponding wavelengths from the equation

$$\lambda = \frac{c}{f} \tag{1.6}$$

where λ is the wavelength, c is the velocity of propagation (3×10^{10} cm/s), and f_o is the sensor operating frequency. From Figure 1.1 power generating systems are of low frequency while radio transmissions occupy a wide band of higher frequencies. Microwave radars are in the GHz frequency region with corresponding wavelengths in centimeters.

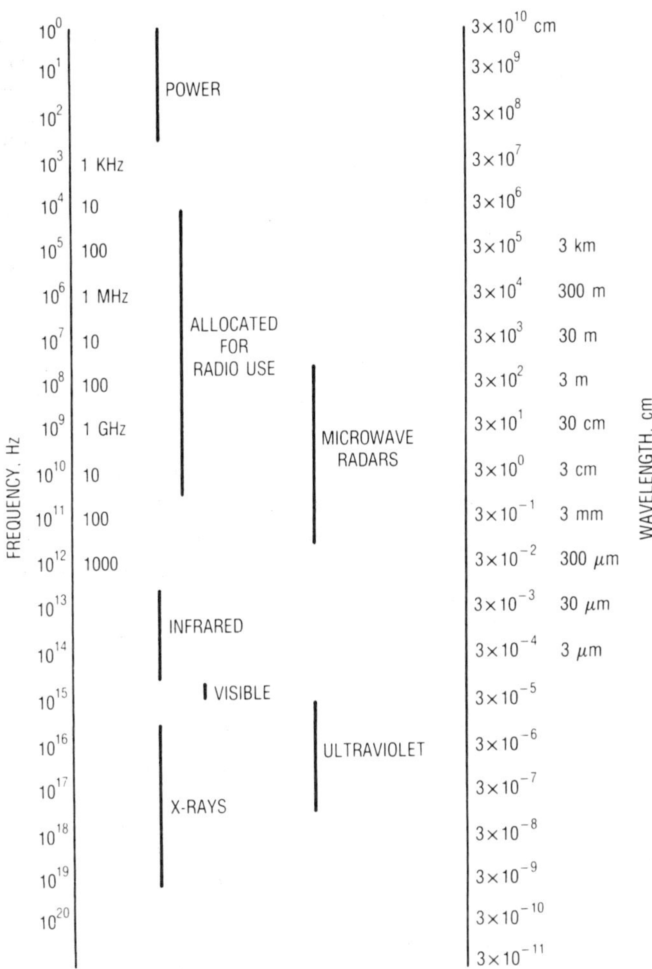

Figure 1.1 Sensor frequencies and wavelengths.

The specific radar frequency bands are given in Table 1.2. Ground-based radars are usually in the L-, S-, and C-band regions, while airborne radars are in the X-band region. MMW radars have frequencies of greater than 30 GHz. The following band designations, V-band 40–75 GHz and W-band 75–110 GHz, are also used in MMW radars. Infrared and EO systems, as shown in Figure 1.1, operate at frequencies orders of magnitude higher than microwave radars.

Table 1.2 Radar Frequency Bands

Band	*Frequency Range*	*Wavelength* (cm)
VHF	30–300 MHz	1000–100
UHF	300–1000 MHz	100–30
L	1000–2000 MHz	30–15
S	2000–4000 MHz	15–7.5
C	4000–8000 MHz	7.5–3.75
X	8000–12,500 MHz	3.75–2.40
K_u	12.5–18 GHz	2.40–1.67
K	18–26.5 GHz	1.67–1.13
K_a	26.5–40 GHz	1.13–0.75
MMW	$f > 30$ GHz	$\lambda < 1.0$

The specific band designations of EO systems are given in Table 1.3. This table also includes the visible region of 0.38–0.76 μm, as well as the ultraviolet region of 0.1–0.38 μm. Electro-optical sensors discussed in this book consist of thermal imaging systems and infrared search and tracking systems; they operate mostly in the 3–5 μm and 8–12 μm wavelength regions. Most laser radars operate at 1.06 μm and 10.6 μm wavelengths.

Table 1.3 Infrared Spectrum

Region	*Wavelength* (μm)
Ultraviolet	0.1–0.38
Visible	0.38–0.76
Very near infrared	0.76–1.0
Near infrared	1–3
Middle infrared	3–8
Longwave infrared	8–14
Far infrared	14–50

1.3. ATMOSPHERIC ATTENUATION

An important consideration in the design of any sensor system is the atmospheric effects on that system. Usually, sensor systems are designed to operate under a given atmospheric condition: rain, fog, and so on. The exception is spaceborne systems, which operate beyond the earth's atmosphere. In this case, the sensors are not affected by atmospheric degradations.

Figure 1.2 shows atmospheric attenuation *versus* frequency for most of the operating sensor systems. Atmospheric attenuation is expressed in dB/km. For example, an attenuation of 0.5 dB/km for a two-way range of 20 km will represent a *signal-to-noise* (S/N) ratio power loss of a factor of 100:

$$
\begin{aligned}
\text{Loss} &= 2 \times 20 \times 0.50 \\
&= 20\ \text{dB} \\
&= 100
\end{aligned}
\tag{1.7}
$$

From Figure 1.2, note that atmospheric attenuation is small at frequencies of less than about 20 GHz. After 30 GHz, there are several atmospheric windows where attenuation is tolerable, at least for short-range applications. These windows are at 35, 94, 140, and 240 GHz. For this reason, MMW radars are built at these frequencies. Observe also that the infrared region has atmospheric windows at 1, 2–5, and 7–14 μm, as will be discussed later.

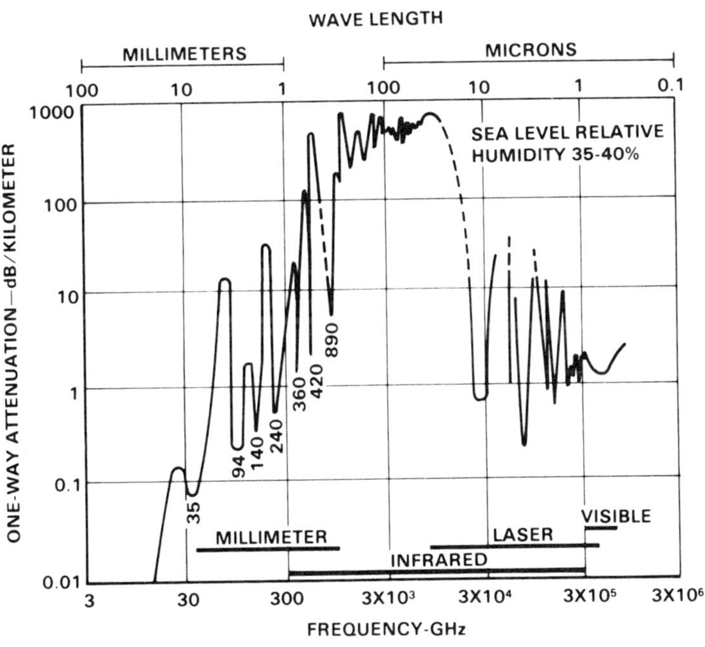

Figure 1.2 Atmospheric attenuation *versus* frequency.

Figure 1.3 shows the approximate atmospheric attenuation values under adverse weather conditions of fog and rain [1, p. 297]. It should be noted that there is considerable uncertainty about these values, as measurements are difficult to obtain under controlled conditions especially for fog and rain. From this figure it is clear that adverse weather conditions considerably reduce the utility of visible and infrared types of sensors and degrade the utility of MMW radars. Microwave radars are not affected so adversely by these weather conditions. Quantitatively, 16 mm/h is considered heavy rain, while 4 and 1 mm/h is considered moderate and light rain, respectively. Figure 1.4 gives a more descriptive presentation of rain and cloud effects on microwave transmission from space. From this figure, note that in the lower GHz region, transmission through weather conditions is practically unaffected.

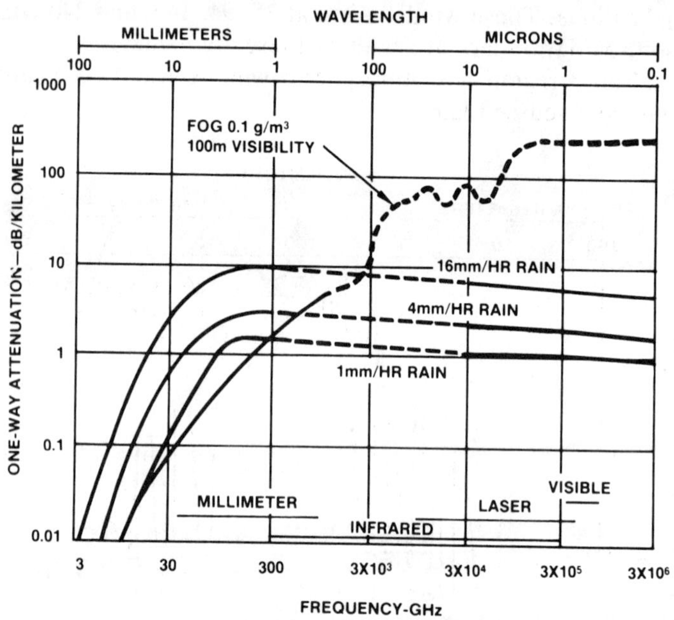

Figure 1.3 Rain and fog attenuation *versus* frequency.

Figures 1.5 and 1.6 give atmospheric attenuation effects on MMW and EO sensor systems in greater detail under clear weather conditions. Figure 1.5 expresses the total atmospheric loss from the surface of the earth to an altitude of 130 km. Note the spread of "low-loss" windows in this figure. Figure 1.6 gives the atmospheric transmission as a function

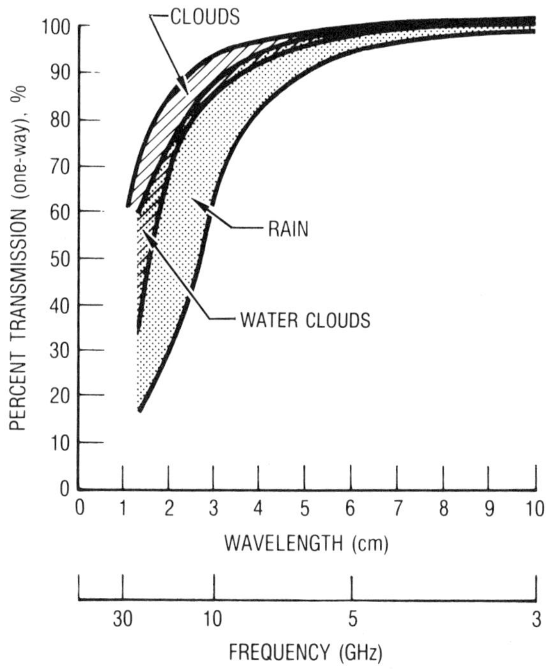

Figure 1.4 Effect of rain and clouds on radio transmission in the spectrum's RF region.

of wavelength for EO systems. Unity in this figure means no atmospheric loss under clear weather conditions. Note the "no loss" region between 0.3–1.0 μm, 3–5 μm, and 7–14 μm. EO and laser radar systems are built to operate in these regions.

1.4. ANGULAR RESOLUTION CONSIDERATIONS

The angular resolution of sensor systems is primarily a function of their beam dispersion characteristic. And, as given in (1.4), this beam dispersion is a function of wavelength λ and aperture dimension D. In cases where accurate angular information is needed, EO systems are utilized. Such is the case of the gun pointing in tank fire control systems. Figure 1.7 gives beam dispersion distance at a range of 10 km for various sensor systems, each having an aperture diameter of 1 ft or about 30 cm. From this figure, note that a microwave radar system at X-band (9 GHz frequency) will have a wavelength of 3.33 cm. From (1.4), this will result in a beamwidth

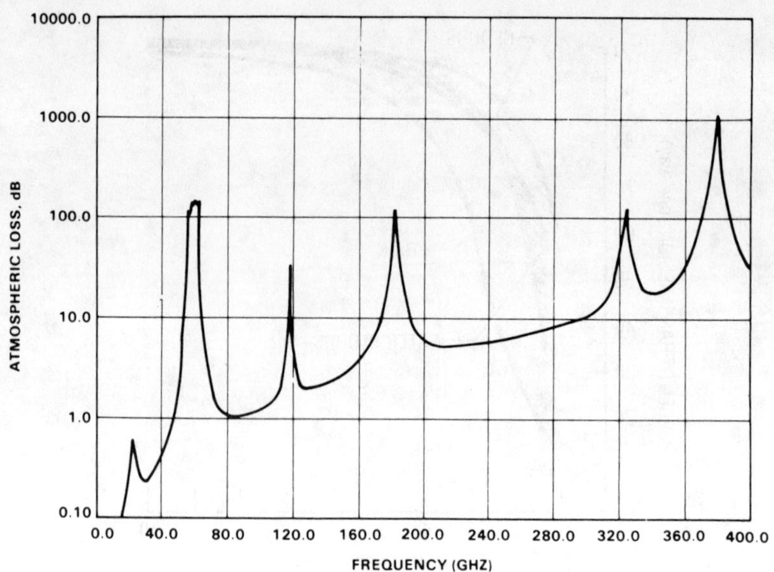

Figure 1.5 Atmospheric loss of MMW frequencies (earth's surface to an altitude of 130 km at vertical incidence).

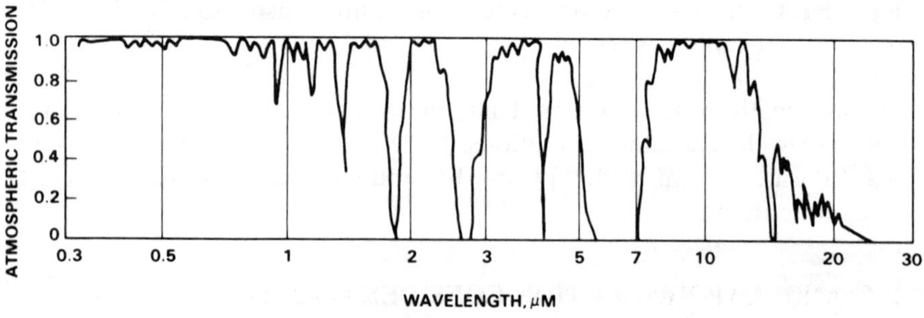

Figure 1.6 Transmission of electromagnetic radiation at μm wavelengths.

of

$$\Theta_{BW} = \frac{3.33}{30}$$

$$= 0.11 \text{ rad}$$

$$= 6.3^\circ \tag{1.8}$$

The corresponding resolution range at 10 km will be

$$\Delta R = R\Theta_{BW}$$
$$= 1100\text{ m} \tag{1.9}$$

Similar calculations for MMW radars, EO thermal imagers, and optical systems will result in resolution cell dimensions of 104 m, 32.8 cm, and 1.64 cm, respectively. Note that the optical systems, because of their low wavelengths, have the highest resolution characteristics. Microwave and MMW radars, because of their large beamwidths, are used for large area searches. In a number of cases, these radars are used as cueing devices for narrow-beam EO systems. The EO systems, because of their high resolutions, can be used for target recognition and identification.

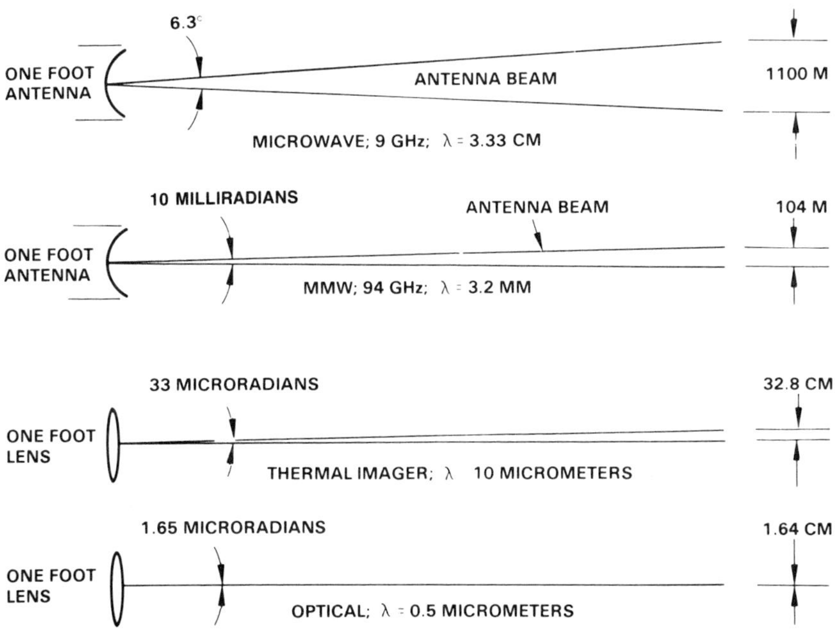

Figure 1.7 Resolution cells at a range of 10 km.

1.5 MICROWAVE RADAR SYSTEMS

Microwave radar systems detect and track targets using two distinct principles: elapsed time between transmitted and received pulses to calculate range, and frequency shift between these pulses to calculate range rate. Figure 1.8 illustrates the timing principle used to calculate radar-target range. A pulse of duration τ is transmitted at the same time that a timer is started to measure the elapsed time between this pulse and the target return pulse. Using this measured time, t, we can calculate the radar-target range:

$$R = \frac{ct}{2} \tag{1.10}$$

where the factor 2 comes from the round-trip distance traveled by the pulse. In (1.10), R is the radar-target range, c is the velocity of propagation in the atmosphere (3×10^{10} cm/s), and t is the elapsed time. The angular position of the target in azimuth and elevation is obtained from the position of the antenna at the time of target detection. Thus, with a known range and angular position, the location of the target can be established. Note that the angular position accuracy of the target will be a function of antenna beamwidth Θ_{BW}.

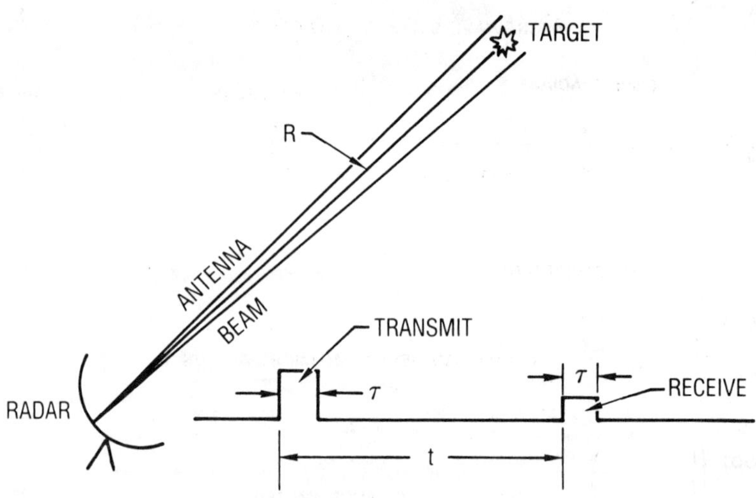

Figure 1.8 Timing of transmitted and received pulses.

Figure 1.9 shows the block diagram of a doppler radar operating on frequency shift of target return signal for detection. The T/R switch of the figure switches the system between transmitting and receiving functions. Note that when the receiver is on, the transmitter is off, and vice versa. The transmitted waveform is of frequency f_o, while the received

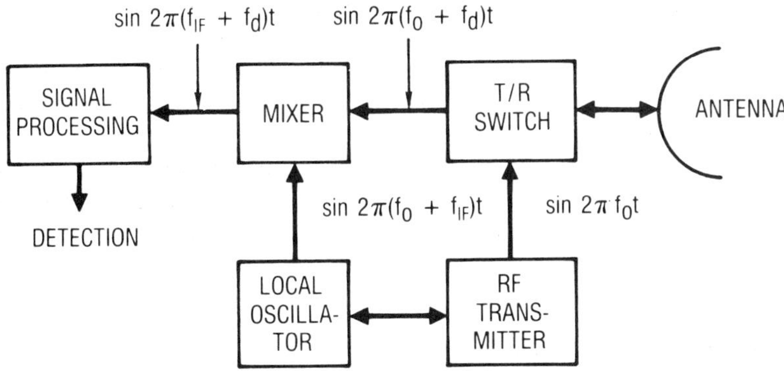

Figure 1.9 Block diagram of a doppler radar.

waveform is at frequency $f_o + f_d$. The doppler shift contained in this waveform is related to target motion as follows:

$$f_d = \frac{2\dot{R}}{\lambda} \tag{1.11}$$

where $\dot{R}$ is the rate of change of R (radar-target range) and λ is the wavelength. Thus, having doppler shift f_d, we can calculate $\dot{R}$.

The received signal is multiplied by the transmitted signal in the mixer of Figure 1.9. This multiplication, usually referred to as a coherent detection (heterodyne), reduces the received frequency from *radio frequency* (RF) to *intermediate frequency* (IF) or lower. The actual value of f_d is obtained by signal processing, which is primarily an implementation of a Fourier transform to obtain the frequency content of the received waveform.

Note that Figure 1.9 shows the return of a single target at f_d. In actual applications, several targets may be present in the antenna beam; this will give rise to several returns, complicating the signal processing load. In a number of modern radars, range (time) and range rate (frequency) processing are combined to obtain both target range and range rate values simultaneously.

Police radars used for detecting speeding automobiles use the doppler principle with a signal processing diagram very similar to Figure 1.9. In these radars, the microwave energy is transmitted through the radar beam (usually a horn antenna) toward the target, which in this case is the suspected speeder. The radar return is multiplied by the transmitted signal to obtain the doppler shift. Having the doppler shift, the relative velocity $\dot{R}$ of the target and the radar can be computed from (1.11). If the radar is stationary $\dot{R}$ will represent the velocity of the target automobile. If not, the velocity of the radar should be subtracted from the measured $\dot{R}$ value.

In a number of radars, as will be discussed in Chapter 2, frequency modulation techniques are used to obtain range and range rate values. These modulation techniques combine the basic time and frequency principles discussed here in the radar signal processors.

1.5.1 Detection Range Calculations

The detection range of microwave radars in the *search mode* of operation is a function of search volume and the radar frame time (i.e., time to scan the search volume) as well as other radar and target parameters. The detection range equation is given as follows:

$$R = \left| \frac{\overline{P}A\sigma}{(16)kTL(\mathrm{S/N})} \cdot \frac{T_s}{\Omega} \right|^{1/4} \tag{1.12}$$

where

R = detection range
$\overline{P}$ = average transmitted power
A = antenna aperture
σ = radar-target cross section
k = Boltzmann constant
T = equivalent receiver noise temperature
L = system losses
S/N = required signal-to-noise ratio for target detection
T_s = radar scan time
Ω = scan volume in steradians

From Eq. (1.12), the radar detection range in the search mode is proportional to the product of transmitted power and the antenna aperture $\overline{P}A$. Additionally, the search detection range of Eq. (1.12) is independent of the frequency. This, however, is not strictly correct, as higher frequencies will result in smaller antenna beams for a given antenna aperture. And, searching a given area with this smaller antenna beam in a TV raster scan, for example, will incur larger search losses. This search loss should be included in the loss term of Eq. (1.12). Note also that Eq.

(1.12) does not include atmospheric losses due to attenuation of electromagnetic energy.

As a numerical example of a search radar, consider the following parameters:

$$\bar{P} = 1 \text{ kW}; \quad A = 4\text{m}^2; \quad \sigma = 20 \text{ m}^2;$$
$$T = 800 \text{ K}; \quad L = 10 \text{ dB } (10); \quad \text{S/N} = 10 \text{ dB } (10);$$
$$T_s = 10 \text{ s}; \quad \Omega = 80° \times 10° \text{ (Az} \times \text{El)} \tag{1.13}$$

Using these radar parameters in Eq. (1.12), we have

$$\begin{aligned} R &= \left| \frac{(1000)(4)(20)(10)}{(16)(1.38\text{E}-23)(800)(10)(10)(80/57.3)(10/57.3)} \right|^{1/4} \\ &= (1.858\text{E}23)^{1/4} \text{ m} \\ &= 656 \text{ km} \end{aligned} \tag{1.14}$$

where E has been used to signify exponent values.

This detection range is for a 20 m^2 radar cross-section target, which is typical of small aircraft and fighters. Note also that the radar cross section, although related to the physical size of the targets, is a function of many other parameters, such as transmit frequency, antenna polarization, and target electromagnetic reflection characteristics.

1.5.2 Tracking Range Computation

Equation (1.12) was given for a search radar where a given volume of space is searched periodically. Microwave radars are also used for continuous tracking of targets such as aircraft and satellites. In this case, a tracking loop is incorporated in the radar to keep the radar on target and to increase positioning accuracy by the integration of returned pulses. The radar tracking-range equation is given as follows:

$$R = \left| \frac{\bar{P} T_i A^2 \sigma}{(4\pi)\lambda^2 (\text{S/N})(kT)L} \right|^{1/4} \tag{1.15}$$

where T_i is the acquisition on-target time, λ is the wavelength, and S/N is the tracking signal-to-noise ratio.

Note that the tracking range of the radar is proportional to average power multiplied by on-target time, which represents the amount of transmitted energy. Additionally, the tracking range depends on transmitted frequency (wavelength) as seen by the term λ^2 in the denominator.

As a numerical example of Eq. (1.15), consider the following parameters of a tracking radar:

$$\begin{aligned} &\overline{P} = 1\ \text{kW}; \quad && T_i = 50\ \text{ms}; \quad && A = 4\text{m}^2; \\ &\sigma = 20\ \text{m}^2; && f_o = 6\ \text{GHz (C-band)}; && \text{S/N} = 3\ \text{dB}; \\ &T = 800\ \text{K}; && L = 10\ \text{dB}(10) && \end{aligned} \tag{1.16}$$

Tracking S/N is usually considerably less than search S/N, because in search radars a higher threshold is used to reduce the number of false alarms. This higher threshold, in turn, requires higher detection S/N ratios. The 3-dB tracking S/N ratio will be equal to a factor 2. The wavelength can be obtained from the given frequency

$$\begin{aligned} \lambda &= \frac{c}{f} \\ &= \frac{3 \times 10^{10}}{6 \times 10^{9}} \\ &= 5\ \text{cm} \end{aligned} \tag{1.17}$$

Using numerical values of Eqs. (1.16) and (1.17) in Eq. (1.15), we get the tracking range of the radar

$$\begin{aligned} R &= \left| \frac{(1000)(50\text{E} - 3)(4)^2(20)}{(4\pi)(5\text{E} - 2)^2(2)(1.38\text{E} - 23)(800)(10)} \right|^{1/4} \\ &= 1232\ \text{km} \end{aligned} \tag{1.18}$$

At this tracking range, the S/N is 3 dB. The tracking loop of the radar usually has a longer time constant than the initial acquisition on-target time of 50 ms. This additional time is used to integrate the S/N and increase angular and range-tracking accuracies.

The tracking ranges of radars are usually longer than the search ranges, as seen in the earlier numerical example. To utilize this additional range, the radar should be initially pointed correctly at the target. This pointing information may come from a priori knowledge of target position or from another sensor, such as an infrared search and track set. With a single microwave radar sensor, however, the target is initially located in the search mode of operation; then it is continuously tracked in the tracking mode to improve its position (azimuth and elevation) and range accuracy. In this case, the effective tracking range will equal the initial search range of the target.

Equations (1.12) and (1.15) are basic to the design *and* performance calculations of radar systems. They apply equally well to MMW radars, as will be seen in section 1.6. These equations, with some minor variations, can also be used in performance calculation of laser radars.

1.5.3 An Interesting Radar Application

In recent years, microwave radars have contributed significantly to our understanding of the environment through production of radar maps and images. In these applications, both timing and frequency processing of radar data are utilized. The x and y direction resolution of these images are obtained through time (pulse duration) and frequency (doppler) processing of radar data, respectively.

Figure 1.10 gives a radar image of a Central American scene obtained by the SEASAT satellite in 1978. The scene shows volcanic towers above a meandering river surrounded by cultivated farmland. The satellite was orbiting the earth at an 800-km altitude. The bright spots of the figure are associated with a concentration of buildings that produce high radar backscattering. The dark areas are *flat* regions with small backscattered energy. The resolution of the figure is 25 m × 25 m and 32 shades of gray were used to produce it. For a complete discussion of this radar application see Chapter 3.

1.6 MILLIMETER-WAVE RADAR SYSTEMS

In general, radar systems with higher frequencies are more difficult and more expensive to produce. Millimeter-wave radars [2] are more expensive than microwave radars. This additional expense is justified by the need for higher angular resolution and a compact size in a number of tactical systems. Similarly, compared to electro-optical systems MMW radars are less sensitive to weather effects and tactical countermeasures used against thermal imaging systems, such as smoke and flares. Additionally, they are degraded to a lesser extent under battlefield conditions of dust and debris.

The wide bandwidth (narrow pulse) allocation afforded by MMW radars results in a fine range resolution which can be used for target imaging or reduction of ground clutter return. Additionally, the high doppler sensitivity of MMW radars affords good resolution of slowly

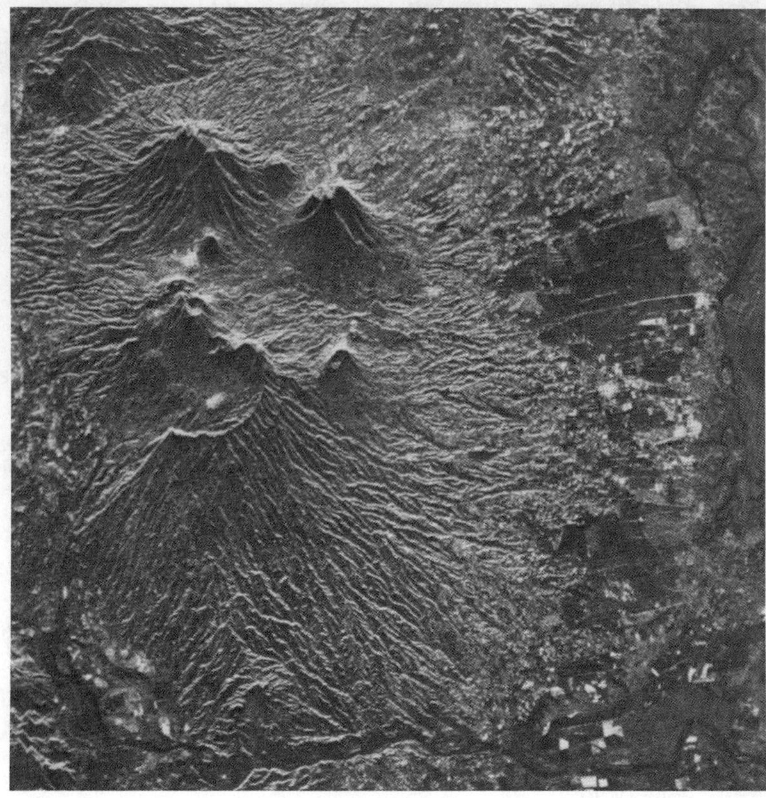

Figure 1.10 Synthetic array radar image of Central America obtained by SEASAT satellite (courtesy of MacDonald Dettwiler and Associates, Ltd.).

moving targets. The range resolution capability of MMW radars is exploited in target recognition studies to obtain range profiles of targets. These range profiles are then compared to a library of known target range profiles for identification [3].

Tactical MMW radar systems are produced for shorter range (10 km) detection of targets. They are attractive because they offer angular resolution, can detect stationary targets, and are less susceptible to electronic countermeasures. As seen from Figure 1.2, MMW radars are produced at frequencies of 35, 94, 140, and 240 GHz because of the low atmospheric attenuation at these frequencies. These higher frequencies, as compared to microwave radars, result in better angular position resolution of targets. Noting that angular resolution and accuracy are a function of antenna beamwidth, θ_{BW}, we have

$$\begin{aligned}\theta_{BW} &= \frac{\lambda}{D} \\ &= \frac{c}{f_o D} \\ &\propto \frac{1}{f_o}\end{aligned} \tag{1.19}$$

From Eq. (1.19), note that beamwidth and thereby angular accuracy is inversely proportional to transmitted frequency. Thus, going from a microwave radar at X-band (9 GHz) to a MMW radar at 94 GHz will improve the angular accuracy by a factor of 10. Additionally, this reduced beamwidth reduces the ground coverage of the radar, which decreases the level of backscattered energy from the ground. This decreased level of ground return allows detection of nonmoving targets on the ground. Another benefit of smaller antenna beam is the reduction of the multipath problem, which occurs when the antenna beam intercepts the ground and the radar energy from the target is bounced off the ground and finds its way to the transmitting antenna. Narrow antenna beams reduce the possibility of multipath interference.

1.6.1 Ground Clutter Radar Cross Section

From Figure 1.11, the ground clutter *radar cross section* (RCS) can be computed by a consideration of MMW radar resolution cell size. At a range of R, the antenna beam Θ_{BW} covers the ground with a circular section of approximate radius $R\Theta_{BW}$. The pulsewidth τ coverage on the ground will be

$$\frac{c\tau}{2\cos\epsilon} \tag{1.20}$$

where c is the velocity of propagation, τ is the pulsewidth, and ϵ is the look angle. Note that the MMW pulsed radar S/N calculations are done on a per pulse basis. Using Eq. (1.20) with antenna beam coverage, we get the equivalent ground RCS

$$\text{RCS} = \sigma_c = (R\Theta_{BW})(c\tau/2\cos\epsilon)\sigma_o \tag{1.21}$$

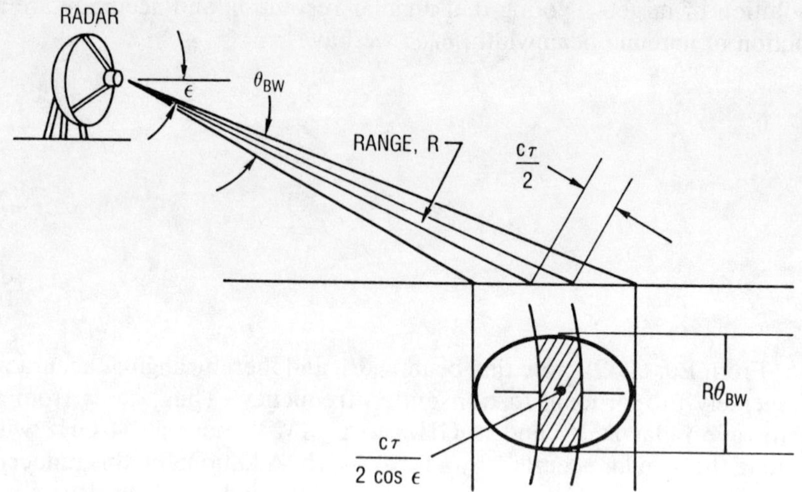

Figure 1.11 Ground clutter RCS.

where σ_c is the equivalent ground clutter RCS. The value of σ_o is the ground backscattering coefficient and is obtained experimentally. This value depends on a number of parameters such as transmitted frequency, polarization, type of ground terrain, and the look angle.

As a numerical example of RCS computation, consider the following parameters:

$$R = 5 \text{ km}; \quad \Theta_{BW} = 0.5°; \quad \tau = 1 \ \mu s$$
$$\epsilon = 10°; \quad \sigma_o = 0.01 \tag{1.22}$$

This value of σ_o for MMW radar frequencies is representative of agricultural fields. Using the parameters of Eq. (1.22) in Eq. (1.21), we get

$$\begin{aligned}\sigma_c &= [(5E3)(0.5/57.3)]\left[\frac{(3E8)(1E-6)}{2\cos 10°}\right](0.01) \\ &= 66.45 \text{ m}^2\end{aligned} \tag{1.23}$$

Thus, each ground resolution cell will have an equivalent target cross section of about 66 m^2.

1.6.2 Rain Backscattering RCS

Since MMW radars are at shorter wavelengths than microwave radars, the backscattering effects of rain should also be considered. The ground clutter and rain backscattering in effect increase the system noise level, as these returns are considerably greater than system noise. Targets should compete with these higher levels of "noise" for detection.

Figure 1.12 illustrates rain RCS calculations. In this case, we are dealing with a volume instead of an area as in the ground clutter calculations. From Figure 1.12, the rain backscattering RCS for a "pulse equivalent" resolution cell is given as

$$\text{RCS} = \sigma_r = \frac{\pi}{4}(R\Theta_{\text{BW}})^2\left(\frac{c\tau}{2}\right)\sigma_i \tag{1.24}$$

where σ_i is the rain backscattering coefficient. As a numerical example of this calculation, consider the following parameters:

$$R = 5 \text{ km}; \qquad \Theta_{\text{BW}} = 0.5°; \qquad \tau = 1\ \mu\text{s};$$
$$\sigma_i = 1.2 \text{ cm}^2/\text{m}^3 \tag{1.25}$$

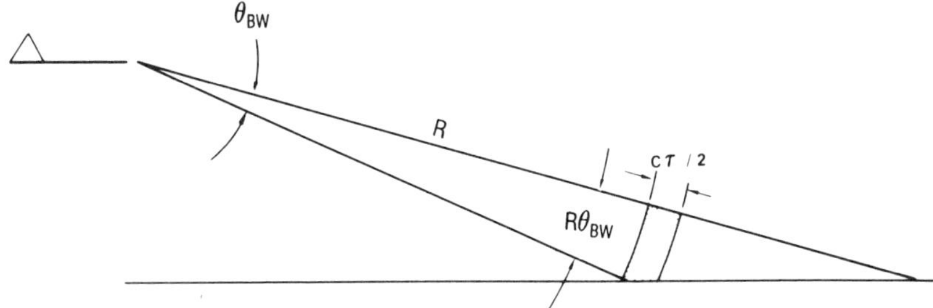

Figure 1.12 Rain backscattering RCS.

In Eq. (1.25), the value of σ_i is representative of the backscattering coefficient at 35 GHz frequency under moderate rain (4 mm/h) conditions. Using the parameters of Eq. (1.25) in Eq. (1.24), we get

$$\sigma_r = \frac{\pi}{4}\left|(5\text{E}3)(0.5/57.3)\right|^2\left(\frac{(3\text{E}8)(1\text{E}-6)}{2}\right)(1.2\text{E}-4)$$
$$= 26.91 \text{ m}^2 \tag{1.26}$$

In addition to backscattering calculation, the atmospheric attenuation losses must also be incorporated in the signal-to-noise ratio computations.

1.6.3 Signal Levels and Detection

In the design and performance calculations of MMW radars, the target, ground clutter, and rain backscattering S/N ratios are computed and plotted as a function of range. Specifying a detection signal-to-noise level, we can compute the detection range from this plot.

In the search mode of operation, the target S/N can be computed from Eq. (1.12) as

$$(\mathrm{S/N})_t = \frac{\overline{P}A\sigma}{(16)kTLR^4}\left(\frac{T_s}{\Omega}\right) \propto \frac{1}{R^4} \tag{1.27}$$

where σ is the RCS of the target. The ground clutter-to-noise ratio can be computed by replacing target cross section σ of Eq. (1.27) by the equivalent clutter cross section of Eq. (1.21). This results in

$$(\mathrm{S/N})_c = \frac{\overline{P}A\Theta_{\mathrm{BW}}c\tau\sigma_o}{(32)(kT)LR^3}\left(\frac{T_s}{\Omega}\right) \propto \frac{1}{R^3} \tag{1.28}$$

Similarly, the rain backscattering signal-to-noise ratio can be computed by replacing σ of Eq. (1.27) by the rain clutter cross section of Eq. (1.24). This results in

$$(\mathrm{S/N})_r = \frac{\overline{P}A\Theta_{\mathrm{BW}}^2c\tau\sigma_i}{128(kT)LR^2}\left(\frac{T_s}{\Omega}\right) \propto \frac{1}{R^2} \tag{1.29}$$

Equations (1.27) through (1.29) give target, ground clutter, and rain signal-to-noise ratios as a function of radar and other parameters. Note that the target signal is proportional to $1/R^4$, while the ground clutter and rain signal levels are proportional to $1/R^3$ and $1/R^2$, respectively.

1.6.4 Computational Procedure

For specified radar and target parameters, computer programs can be written to compute target, clutter, and rain backscattering S/N as a function of range, including atmospheric attenuation. The atmospheric attenuation effects are included by graphical or analytical methods as will be discussed in future chapters. The results of one such calculation are given in Figure 1.13. From this figure, the target detection range can be obtained for any given signal-to-noise ratio. For example, a target-to-clutter signal power requirement of 10 dB will result, from Figure 1.13, at a detection range of approximately 1.4 km. By specifying a detection signal-to-noise ratio, the procedure depicted in Figure 1.13 can be programmed for computer solution.

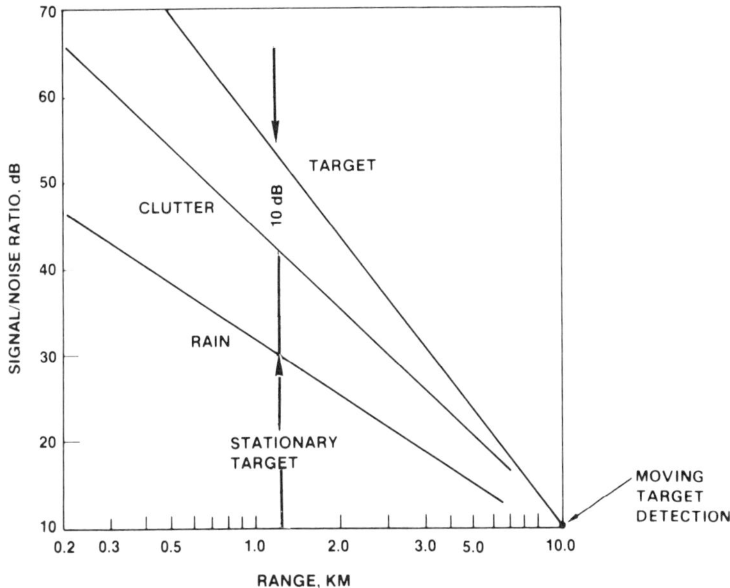

Figure 1.13 Typical S/N *versus* range figures.

Note that moving targets can be detected by doppler processing; they can be separated from ground and rain returns because of the latter's stationary nature. See Chapter 2 for a complete discussion of doppler processing. Thus, moving targets are detected against system noise levels rather than clutter levels. Assuming that a 10 dB S/N is sufficient for detection of moving targets, from Figure 1.13, these targets can be detected at a range of 10 km.

Figure 1.14 gives performance results of two typical MMW radars operating at frequencies of 35 and 95 GHz. The specific parameters of these radars are given in Table 1.4. From Figure 1.14, note that moving targets can be detected at longer ranges by 35-GHz radars than by 95-GHz radars. Stationary targets, however, are detected at longer ranges

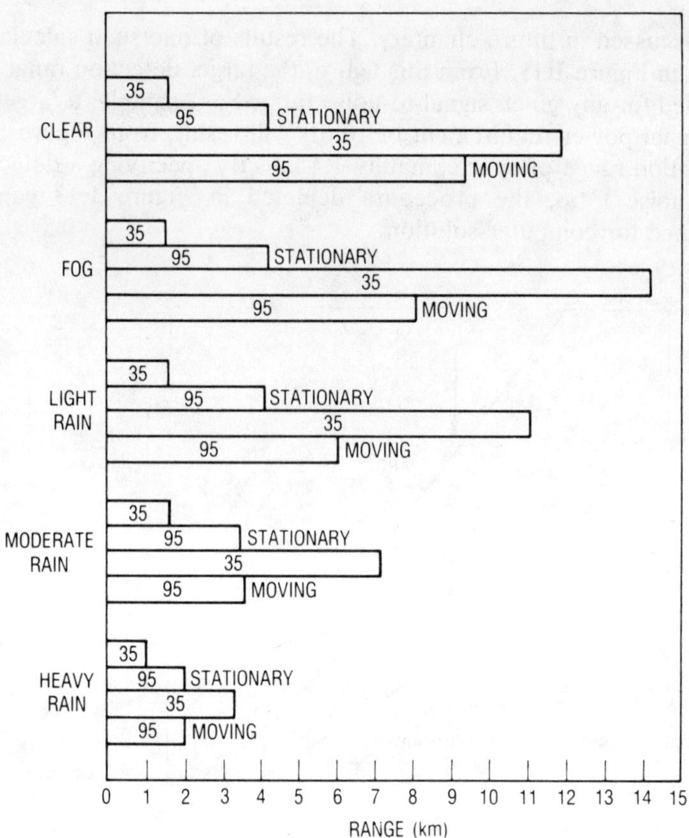

Figure 1.14 Detection ranges of stationary and moving ground targets for 35 and 95 GHz MMW radars for a 1-foot parabolic antenna.

by 95-GHz radars, primarily because of the smaller beamwidth and the resulting smaller clutter region of these radars.

1.7 ELECTRO-OPTICAL THERMAL IMAGERS

The EO thermal imaging systems [4–7] are passive devices designed to operate at wavelengths of 3–5 and 8–14 μm, where there are atmospheric transmission windows, as discussed previously and illustrated in Figure 1.6. Electro-optical thermal imagers are also referred to as *forward-looking infrared systems* (FLIR) and night vision systems, because of their ability to produce images at night. These systems consist of five parts: optical lens, scanners, detectors, signal processing, and displays, as illustrated in Figure 1.15. Electromagnetic radiation from the source is collected by the optical lens and channeled to the detector, where it is converted to electrical energy. The conversion is done in the detector either by changes in resistivity (photoconductive) or by generation of electric current (photovoltaic). The electrical signal is processed to enhance the S/N ratio and to define the target position within a given reference frame.

Table 1.4 MMW Radar Parameters of Example

Item	*Symbol*	*35 GHz*	*95 GHz*
Peak power (W)	P	20	20
Pulse width (s)	τ	50E − 9	50E − 9
PRF (Hz)	PRF	10,000	10,000
RCS (m^2)	σ	30	30
Antenna aperture (m^2) (12-in diameter)	A	0.073	0.073
Radar frame time (s)	t_s	3.6	4.1
Equivalent noise temp. (K)	T	1170	1470
Losses (dB)	L	9	13
Azimuth scan (deg)	Az	60	60
Elevation scan (deg)	El	4	4
Antenna beamwidth (deg)	Θ_{BW}	1.60	0.60
Ground backscattering coefficient	σ_o	0.01	0.01

Figure 1.15 shows an EO imaging system with a single detector. The object space resolution cell is imaged on the detector through horizontal and vertical scanners. Rotation of the horizontal scanner effectively moves the detected resolution cell across the horizontal object space, while the vertical scanner moves this point in the vertical plane of the object space. Thus, the movement of the scanners puts the entire scene through the single detector, producing the appropriate image on the display. In case of multiple detector systems, more than a single (in object space) resolution cell can be mapped into the detectors. This eventually increases the signal processing S/N ratio of each resolution cell, enhancing the quality of the resulting thermal image. In *focal plane arrays* (FPA), a mosaic of detectors is used to map an entire scene. This is similar to photographic cameras, where the lens' field of view is reproduced on the film. Although work is continuing toward entire scene registration using FPA concepts, it may be some time before these systems become operational.

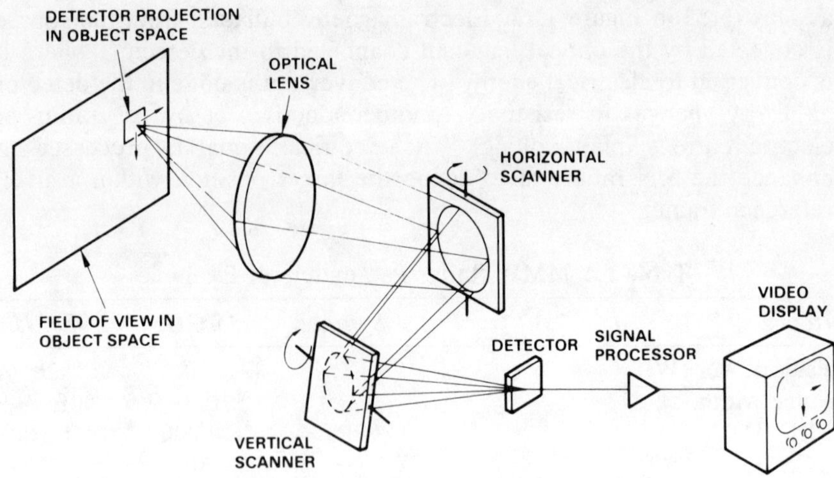

Figure 1.15 Electro-optical sensors.

Infrared detectors of lead sulphide, lead selenide, and lead telluride achieve acceptable responsivity levels in an uncooled environment. Lead sulphide is used for detection in 1.5–3.0 μm band, and lead selenide is used in the 3–5 μm band. The sensitivity of these detectors increases in the cooler environment. Therefore, the development of detector materials was paralleled by development in cooling methods. Thermoelectric coolers were developed to cool the detectors to 200 K. Following this, a major breakthrough was made by Joule-Thompson cryogenic coolers, achieving temperatures of 70–80 K. This allowed the use of the mercury-cadmium-telluride family of detectors to operate in the 8–12 μm region of the infrared spectrum.

1.7.1 Performance Calculations

The performance of EO systems is a measure of their ability to detect targets in a ground scene setting; for example, a tank or a truck on a road surrounded by agricultural fields. The EO systems produce a TV-like picture of the scene by utilizing temperature gradients across the object space. Since these systems work on temperature variation, as opposed to light illumination of photographic systems, the EO images can also be produced at night.

The quantitative performance of the EO thermal imaging system is measured by obtaining a curve of *minimum resolvable temperature* (MRT) as a function of spatial frequency (range). The equation for MRT, derived in Chapter 5, is given as follows:

$$\text{MRT} = \frac{24(f/\#)}{\pi\tau_a D}\left(\frac{f_s}{r_s}\right)\left(\frac{1}{K_n}\right)^{1/2}\left(\frac{\Omega_s}{nT_f\alpha_d}\right)^{1/2}\left(\frac{1}{T_e\dot{F}}\right)^{1/2} \cdot \frac{1}{\int_{\lambda_1}^{\lambda_2}\frac{\partial W(\lambda,T)}{\partial T}D^*(\lambda)d\lambda} \tag{1.30}$$

where

MRT = minimum resolvable temperature
$f/\#$ = focal number
τ_a = optical efficiency
D = aperture diameter
f_s = spatial frequency
r_s = modulation transfer function
K_n = scanning efficiency
Ω_s = scan volume
n = number of detectors
T_f = scan time
α_d = detector field of view
T_e = eye integration time for scope display detection
$\dot{F}$ = frame rate
$\int_{\lambda_1}^{\lambda_2}$ = spectral emittance of background
D^* = detector detectivity

Most of the parameters of Eq. (1.30), as just described, are self-explanatory. Some of these parameters, however, are defined specifically for EO systems to facilitate design and performance calculations.

The spatial frequency, f_s, expressed in cycles per milliradians, is used to convert the size of the target to a detection or recognition range. For a 1 m dimension target, one spatial frequency cycle represents 1 km of detection range.

The modulation transfer function, r_s, combines various system efficiency factors such as optics, electronics, detector, monitor, and so on. The *modulation transfer function* (MTF) has a value of unity (100 percent) at zero spatial frequency (dc level) and decreases from this value with increasing frequency. The cutoff frequency is usually determined by the lens diffraction characteristic.

The eye integration time, T_e, is used to accumulate target signals as they are painted on the display scope. This experimentally obtained value is ~0.2 s. Thus, if the thermal imager picture is painted on the display scope at a rate of 30 times/s, eye integration time will be equivalent to six frames of signal integration.

The integral term in the denominator of Eq. (1.30) is a measure of the background characteristics on which the target is located. In the following numerical example, the spectral emittance of a grass-covered field is used in the computation of this integral.

The parameter D^* is related to the detector sensitivity. Detector sensitivity, as well as background emittance characteristics, is a function of the wavelength, λ. For this reason, $D^*(\lambda)$ appears within the integral. The maximum detectivity values are usually given to specify detector sensitivity characteristics.

1.7.2 Numerical Example of MRT Calculation

As a numerical example of MRT calculations, consider the parameters of the thermal imaging system given in Table 1.5. This system operates in the 8–12 μm band with a lens 8 in. in diameter. The field of view is 2° × 1.0°. The number of detectors is 1024, and the object space is scanned at a rate of 30 frames/s, resulting in a frame time of 1/30 = 3.34E − 2 s. The detector instantaneous field of view is 50 μrad in each direction (azimuth and elevation). Peak detector detectivity is given at 2E11 cm $\text{Hz}^{1/2}$/W, and the optical efficiency is 0.5.

In addition to the parameters of Table 1.5, the integral term of Eq. (1.30) is evaluated for a normalized detectivity, with a peak detectivity value of unity:

$$\int_{8}^{12} \frac{\partial W[\lambda,T]}{\partial T} D^*(\lambda) d\lambda = 8.4 \times 10^{-5} \text{Hz}^{1/2}/\text{cm sr K} \tag{1.31}$$

Using the parameters of Table 1.5, the integral of Eq. (1.31), and appropriate MTR characteristics, in Eq. (1.30), we can calculate the MRT values as a function of spatial frequency as given in Figure 1.16.

In Figure 1.16, MRT values are proportional to S/N ratios, while spatial frequency values are proportional to range. This figure shows that lower temperatures (low S/N ratios) are detected at shorter ranges, while higher temperatures are detected at longer ranges. The extreme value of spatial frequency corresponds to the *null* point of the optical lens and is given by

$$\bar{f}_s = \frac{1}{(\lambda/D)(1000)}$$

$$= \frac{1}{(10E - 6)/(8 \times 2.54\,E - 2)(1000)}$$

$$\simeq 20 \text{ cycles/mr} \tag{1.32}$$

where λ/D is expressed in mr. The center point of the wavelength is used in Eq. (1.32); that is, 10 μm.

Table 1.5 Parameters of the Thermal Imaging System Example

Spectral band		=	8–12 μm
Optical diameter	D	=	8 in.
Focal number	$(f/\#)$	=	2
Scanning efficiency	K_n	=	0.8
Sensor field of view		=	2.0° (Az)
			1.0° (El)
Number of detectors	n	=	1024
Frame time	T_f	=	3.34×10^{-2} s
Detector instantaneous field of view	$\alpha_d^{1/2}$	=	50 μrad
Frame rate	$\dot{F}$	=	30/s
Eye integration time	T_e	=	0.2 s
Peak detectivity	D^*	=	2×10^{11} cm $\mathrm{Hz}^{1/2}$/W
Optical efficiency	τ_a	=	0.5

Using the MRT curve of Figure 1.16, we can compute the detection range of various targets. For example, consider a tank target with a minimum dimension of 3 m and a temperature differential of 3° from the background. From Figure 1.16, 3° corresponds to a spatial frequency of 18 c/mr. Multiplying this value by the dimension of the tank will result in a detection range, R, of

$$R = 3 \times 18 = 54 \text{ km} \tag{1.33}$$

Note that this detection range does not include atmospheric attenuation effects. Even nominal weather effects will reduce this value to about 25 km.

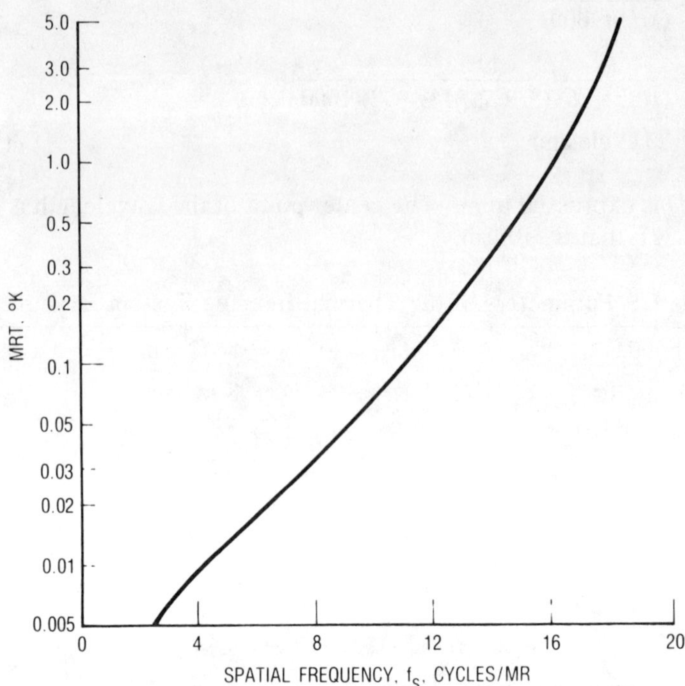

Figure 1.16 MRT *versus* spatial frequency for numerical example.

1.7.3 A Typical Thermal Imager Picture

Figure 1.17 gives a typical thermal imager picture taken of a city street, by an imager operating in the 8–12 μm wavelength. Note that this TV-like picture can be produced at night as well as during the day, since it is produced from temperature gradients rather than light reflections. Note that areas of higher temperatures such as radiator grills and tires appear brighter than colder surfaces, such as plate glass and headlights. Trees and natural objects are somewhere in between.

1.8 LASER RADAR SYSTEMS

Laser radar systems [8, 9] are active devices that operate similarly to microwave radars but at a much higher frequency. This higher frequency has the beneficial effect of smaller components and remarkable angular resolution. The atmospheric attenuation losses, however, are considerable

Figure 1.17 Example of a thermal imager picture .

at these high frequencies. Laser systems built to operate on the ground are usually of limited range (about 10 km), primarily because of atmospheric attenuation effects. Spaceborne laser systems, however, can be built with ranges in the thousands of kilometers in the absence of atmospheric attenuation.

Laser radars, primarily because of atmospheric windows (Figure 1.6) and the availability of detectors are constructed at 1.06 and 10.6 μm wavelengths. The 1.06-μm laser radar systems are referred to as neodymium YAG crystal lasers and 10.6 μm laser systems are called CO_2 gas lasers. The solid-state laser systems at 1.06 μm (1 percent efficiency) are not as efficient as the CO_2 lasers at 10.6 μm (10 percent efficiency).

1.8.1 Performance Calculations

The target signal power received from a laser radar system can be derived from physical principles in the following form:

$$S = \underset{①}{\frac{P}{4\pi R^2}} \cdot \underset{②}{\frac{4\pi}{\frac{\pi}{4}\Theta_{\mathrm{BW}}^2}} \cdot \underset{③}{\frac{\pi}{4}(R\Theta_{\mathrm{BW}})^2} \cdot \underset{④}{\rho} \cdot \underset{⑤}{\frac{1}{2\pi R^2}} \cdot \underset{⑥}{A} \cdot \underset{⑦}{\tau_o} \qquad (1.34)$$

where

S = target signal power at the laser radar

R = radar-target range
P = transmitted pulse power
Θ_{BW} = beamwidth of the laser beam
ρ = backscattering coefficient of the target
A = aperture area
τ_o = optical efficiency

The first term of Eq. (1.34) represents the distribution of power over a sphere of radius R. The second term is the gain of the optical lens where the power is concentrated in a beam of Θ_{BW}^2 steradian instead of a sphere of 4π steradians. The third term represents the area of the target that, in this derivation, is assumed to be larger than the laser beam. This assumption should be removed if the laser beam is larger than the target at range R. The fourth term is the backscattering coefficient of the target. The fifth term is the mathematical expression of the fact that the transmitted power is distributed on a semisphere of radius R after interception with the target. Another, often-used assumption for this distribution is that of Lambertian, where the power is distributed over π instead of 2π steradians. This returned power from the target is intercepted by the aperture area of the lens A as shown by the sixth term. The seventh term of Eq. (1.34) represents the optical efficiency of the laser radar system.

Equation (1.34) can be simplified to this form:

$$S = \frac{P\tau_o\rho A}{2\pi R^2} \tag{1.35}$$

If the laser beam does not subtend the target as expressed by the third term of Eq. (1.34), this term should be substituted by laser radar RCS of the target. Denoting this RCS by σ, we get

$$S = \frac{8P\sigma\rho A^2\tau_o}{\pi^3 R^4\lambda^2} \tag{1.36}$$

In this derivation, the laser beamwidth of Eq. (1.4) was used. Furthermore, assuming a circular lens, the lens area of A was substituted by $\pi D^2/4$.

Comparison of Eqs. (1.35) and 1.36) shows that in cases where the laser beam is larger than the target, the signal is inversely proportional to the fourth power of range. In cases where the beam is smaller than the target, the signal level is inversely proportional to the square of laser-target range.

1.8.2 Laser Radar Receivers

Laser radar receivers are of two types: direct and heterodyne. Direct receivers are similar to EO detectors, where they detect the backscattered energy from the target. Heterodyne receivers contain stable local oscillators, where the transmitted waveform is multiplied by the target return waveform. This allows the reduction of laser received energy to lower frequency levels, where it can be amplified to enhance the return. Figure 1.18 shows a block diagram of direct and heterodyne laser radar receivers. Note that in the heterodyne receiver the transmitted waveform, generated by the CW source, is used in the receiver to reduce the frequency of the received signal prior to the detector.

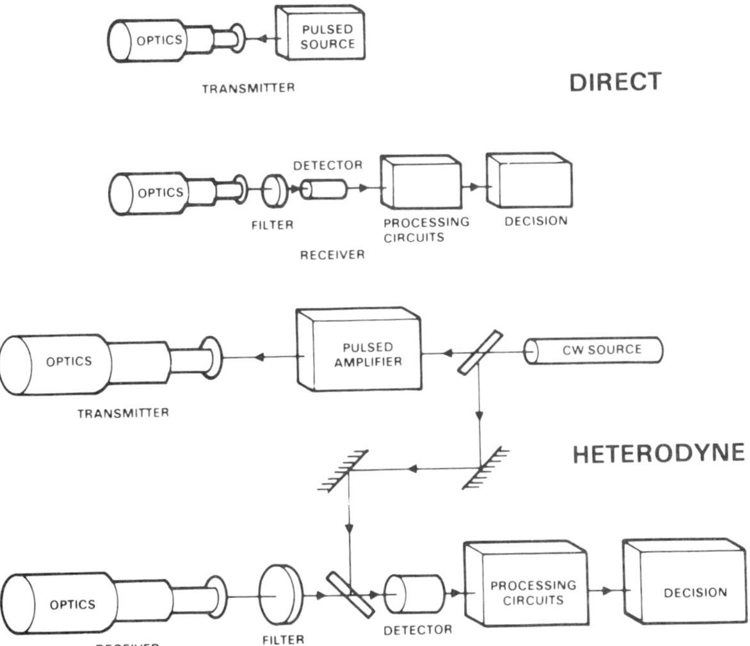

Figure 1.18 Laser receivers.

The detector noise level for *direct* detection is given as

$$N = (A_d \Delta f)^{1/2}/D^* \qquad (1.37)$$

where

N = noise level
A_d = detector area
Δf = receiver bandwidth
D^* = detector detectivity

Similarly, the noise level of *heterodyne* receivers is given as

$$N = (hf/\eta)(B) \tag{1.38}$$

where

N = noise level
h = Planck's constant
f = transmit frequency
η = quantum efficiency
B = receiver bandwidth

Heterodyne receivers are usually more sensitive, but their construction is complex as the coherency of the signal should be preserved for the heterodyning (mixing) process.

1.8.3 Numerical Example

Consider the following parameters for a spaceborne (negligible atmospheric attenuation) CO_2 (10.6 μm) heterodyne laser radar system:

$$P = 20E3\ \text{W}; \quad \tau_o = 0.5; \quad \rho = 0.1; \quad \sigma = 0.2\ \text{m}^2$$
$$D = 50\ \text{cm}; \quad \tau = 100\ \text{ns}; \quad \lambda = 10.6\ \mu\text{m}; \tag{1.39}$$

Let us calculate the S/N at a range of 200 km. The lens diameter of 50 cm will give an aperture area of

$$A = \frac{\pi}{4} D^2$$
$$= 0.196\ \text{m}^2 \tag{1.40}$$

Using the values of Eqs. (1.39) and (1.40) in Eq. (1.36) will result in

$$S = \frac{8(20E3)(0.2)(0.1)(0.196)^2(0.5)}{\pi^3(200)^4(1000)^4(10.6E-6)^2}$$
$$= 1.102E - 11\ W$$
$$\simeq -109.5\ dB \tag{1.41}$$

The noise level of the system can be completed from Eq. (1.38). The transmitted frequency is

$$f = \frac{c}{\lambda}$$
$$= 2.83\ E13\ Hz \tag{1.42}$$

The receiver bandwidth is assumed to be matched to the pulse width; that is:

$$B = \frac{1}{\tau}$$
$$= 10E6\ Hz \tag{1.43}$$

Using the values of Eqs. (1.42) and (1.43) together with a quantum efficiency of 0.5 in Eq. (1.38), we get

$$N = [(6.6256E - 34)(2.83E13)/0.5](10E6)$$
$$= 3.75\ E - 13$$
$$= -124.25\ dB \tag{1.44}$$

The S/N ratio from Eqs. (1.41) and (1.44) will be

$$S/N = 14.75\ dB \tag{1.45}$$

This S/N is considered sufficient for target detection.

Assume that the *direct* detection process, instead of the heterodyne process, is implemented in the radar of the numerical example. For this, consider a detector area of

$$A_d = 4.5\ E - 10\ m^2 \tag{1.46}$$

and detectivity D^* of 1E11 cm – $Hz^{1/2}/W$. Using these values and the bandwidth of Eq. (1.43) in Eq. (1.37), we get

$$N = [(4.5\,E - 10)(10E6)]^{1/2}/1E11$$
$$= 6.70\,E - 13$$
$$= -121.7\ \text{dB} \tag{1.47}$$

The noise level of Eq. (1.47), together with the signal level of Eq. (1.41), will result in a signal-to-noise ratio of

$$S/N = 12.2\ \text{dB} \tag{1.48}$$

This S/N is slightly less than the heterodyne value expressed in Eq. (1.45). Note also that heterodyne laser radar signal processing, in addition to target-radar range, will provide the radar system with radar-target range rate.

1.8.4 Laser Radar Application

In addition to finding range and range rate of targets in the detection and tracking mode of operation, laser radars are used in imaging objects or areas of the ground. In this mode of operation, because of their small beamwidths (high angular resolution) and accurate range values, these radars are capable of producing isometric pictures of targets.

As an example, Figure 1.19 gives a picture of a tank produced by a 10.6 laser radar with an angular resolution of 100 μr and a millimeter wave radar with angular resolution of 800 μr. In both cases, the radar beam was scanned across the tank, and signal strengths were measured to produce this picture. Note the number of circular resolution cells in the MMW radar image. The laser radar picture, because of very small resolution cell size, has an almost photographic appearance.

Another set of thermal imager and laser radar pictures is given in Figure 1.20. This figure shows a picture of an industrial scene for visible, passive thermal imager (8–12 μm) and active laser radars of 10.6 and 1.06 μm wavelengths. The visible wavelength region photograph is the most clear and the most familiar presentation of the scene. The passive thermal imager picture is almost as good. Variations exist, however, between these imager pictures, as the registered signal power of each resolution cell in one case is that of visible light energy and in the other case is that of thermal energy.

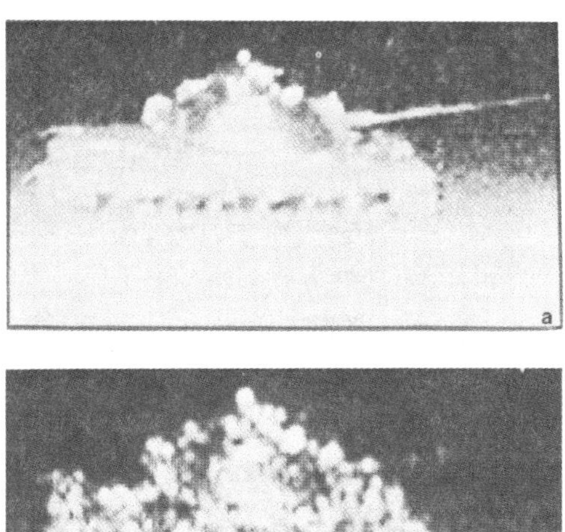

Figure 1.19 Laser (a) and 95-GHz MMW (b) pictures (from [10]).

The active laser radar (10.6 μm) is almost at the mid-wavelength of the passive (8–12 μm) region. Note that this picture, because of missing wavelength regions, is not as clear as the passive picture. The active laser radar of 1.06 μm produces a picture that registers signal powers reflected at this wavelength.

1.9 EXAMPLES OF SENSOR SYSTEMS

Microwave radar systems are designed for a variety of applications, which include remote sensing of extraterrestrial objects, search and detection of targets within the earth's atmosphere as well as outer space, and a variety of military and civilian applications. Figure 1.21 shows a typical shipboard search and target tracking antenna used in conjunction with a radar set. The feed horn of the antenna is supported by the tripod in front of the parabolic dish. The transmitted power through the feed is reflected by the antenna dish to produce the antenna beam. In the search mode of operation the antenna is physically moved in a TV raster scan to cover the desired search volume. The conical pencil beam of the antenna covers the entire search volume much like a scanning search

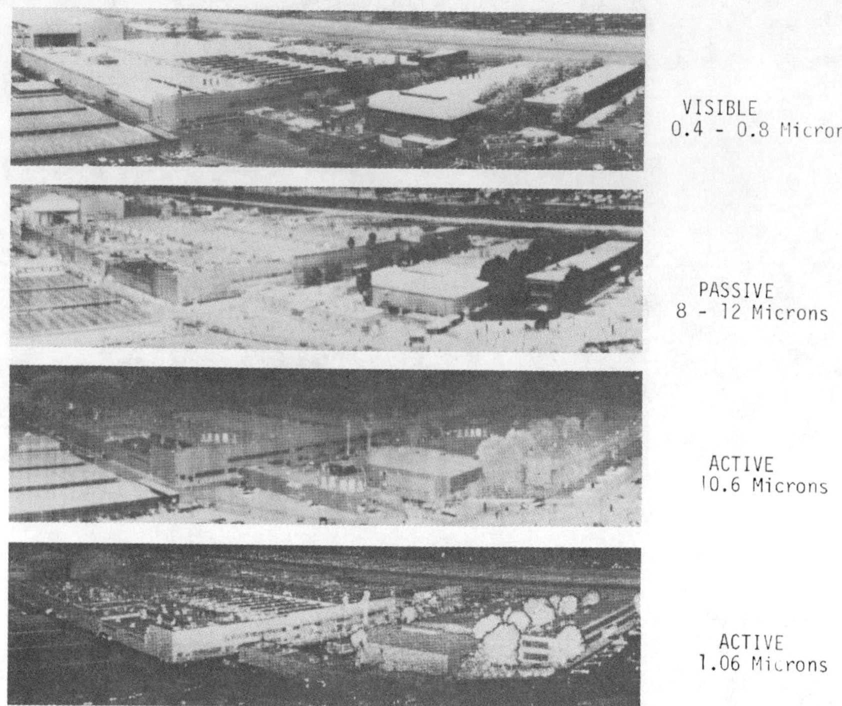

Figure 1.20 Thermal image pictures of an industrial scene (courtesy of Hughes Aircraft Company).

light. In the single target tracking mode, the antenna beam is locked on a single target, following that target and gathering angular position and range information.

Figure 1.22 shows a very large electronically scanned antenna used for detection of ballistic missiles at long ranges. In this case the antenna beam, both in search and tracking modes of operation, is scanned electronically without any physical motion of the antenna. The spots on the face of the antenna are radiating elements controlled by computers to produce and direct the antenna beam in a given direction.

Figure 1.23 shows a mockup of a Ferranti Defense Systems airborne radar system designed and built for the European fighter aircraft. This radar has air-to-air target search and tracking, as well as air-to-surface mapping capability. The antenna is a planar array design with a flat surface. The waveguide structure in the back of the antenna is used for radiation feed. In Figure 1.23, the radome is opened to the left and the radar set is pulled out for inspection.

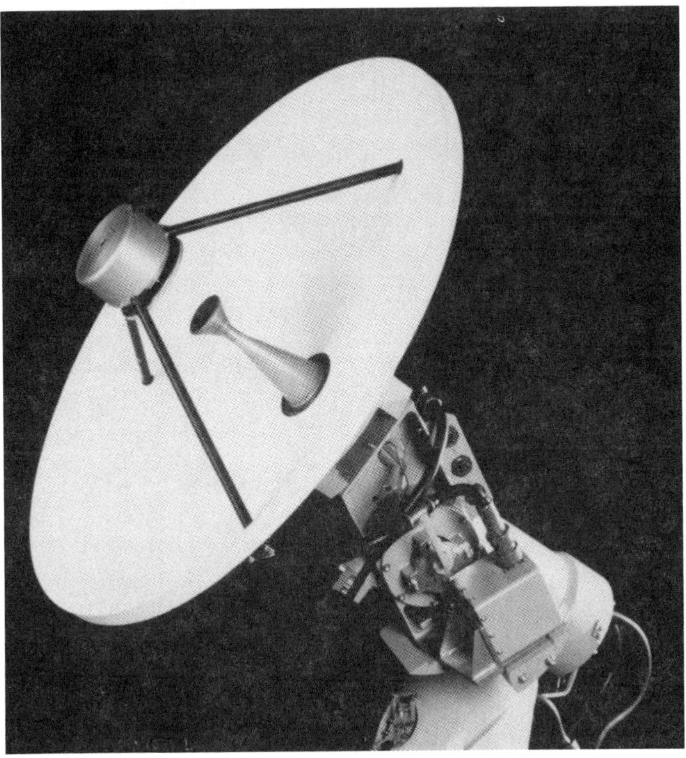

Figure 1.21 Shipboard search and tracking antenna (courtesy of Datron Company).

Figure 1.24 shows a typical radar display of a ground-based radar. The large circular display shows target information after signal and computer processing. The gridlines on this display designate distances from the radar at the center of the display. Control buttons are located on both sides of the display to command the antenna and produce target information desired by the operator. The upper "square display screen" is used for raw data display.

Figure 1.25 shows an infrared tactical imaging system, also called *forward looking infrared* (FLIR) and night vision system. The round window in front of the sensor is used to protect the optical lens structure in the back. The sensor is capable of moving in the vertical plane and the entire structure can be moved in the horizontal plane, for full horizontal and vertical coverage. These thermal imagers are used for navigation and target detection and recognition at night where optical means are degraded.

Figure 1.22 Electronically scanned antenna for detection and tracking of targets at long ranges (courtesy of Raytheon Company).

Figure 1.26 shows several applications of thermal imaging systems. In *wide field of view* (WFOV) of 30 × 40 degrees, the system can be used for navigation. In medium and narrow fields of view, the system can be used for detection and recognition, respectively. In going from a wide to medium to narrow *field of view* (FOV), the operation of the sensor is similar to that of the zoom lens of photographic or TV cameras, where a portion of the original FOV is expanded into the full viewing screen.

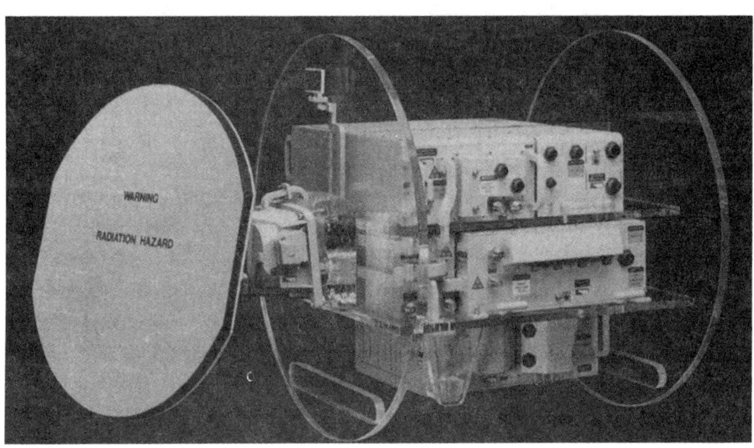

Figure 1.23 Mockup of an airborne radar system with planar array antenna (courtesy of Ferranti Defense Systems).

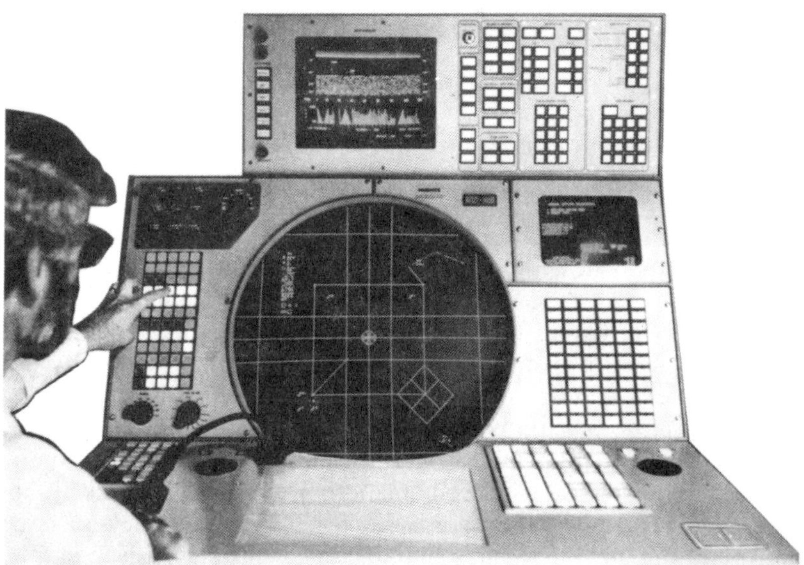

Figure 1.24 Typical ground based radar controls and displays.

Optical and infrared sensors are used to produce multispectral pictures both from airplanes and spacecrafts. Various wavelengths are

Figure 1.25 Airborne thermal imager system mounted on the bottom of aircraft.

used in producing these pictures, with each wavelength region highlighting specific characteristics of mapped areas. Optical spectral bands of 0.4–0.8 μm are generally used for coastal water and vegetation mapping while infrared bands of 0.8–15 μm are used for measuring thermal contents. Figure 1.27 shows a picture of thematic multispectral mapper used in conjunction with Landsat spacecrafts.

Figure 1.28 shows typical thematic maps of a 5 × 5 km region of Mooswald, Federal Republic of Germany, with 30 × 30 m resolution cells. Different spectral band combinations were used to produce these maps. Comparison of these figures shows that agricultural and settled areas are essentially the same in both pictures. The comparison of forest areas, however, will result in differentiation of types, as well as texture, of trees.

Figure 1.29 shows basic elements of a laser for the generation of the laser radar pulse. The center portion consists of a cavity that resembles a fluorescent light bulb. On one end of the cavity is a fully reflecting mirror; on the other end is a partially reflecting mirror. Coupled between

Figure 1.26 Simulated thermal imager display for wide, medium, and narrow fields of view.

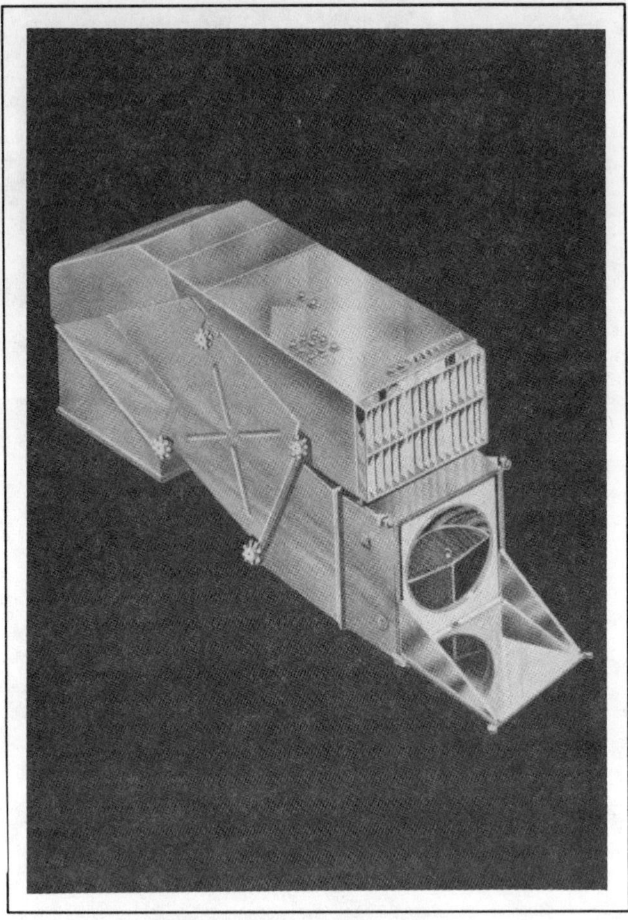

Figure 1.27 Thematic mapper for multispectral mapping and remote sensing of earth resources.

them is the active medium (a ruby rod for ruby laser and a gas rod for the CO_2 gas laser) or a semiconducting material such as gallium arsenide. Energy is pumped into this active medium, causing excitation of the molecules, which form a narrow coherent laser beam.

Figure 1.30 shows the production of an experimental laser beam. This tunable solid state laser can be used in satellite communication and remote sensing.

Figure 1.28 Multispectral thematic maps of Mooswald region of West Germany (DFVLR, Federal Republic of Germany).

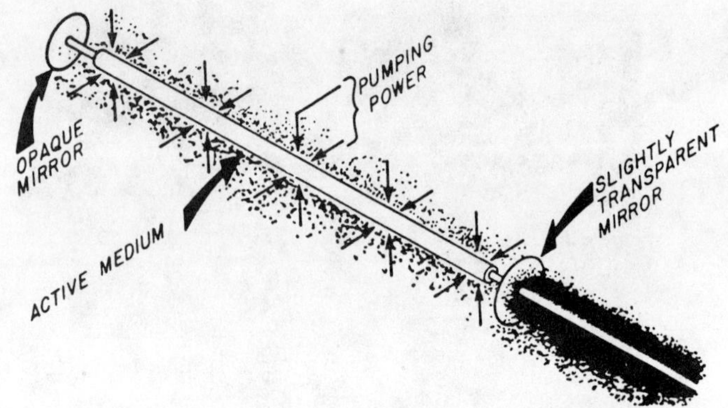

Figure 1.29 Basic laser elements (from [9]).

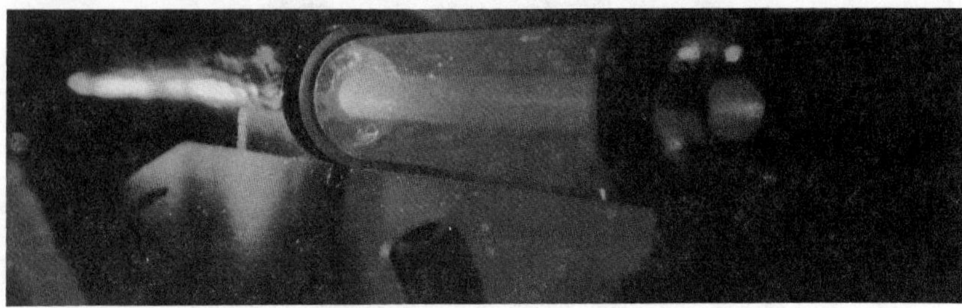

Figure 1.30 Generation of laser beam in an experimental laser system.

1.10 COMPARISON OF MICROWAVE AND ELECTRO-OPTICAL SYSTEMS

Electro-optical thermal imager systems operate in the passive mode by utilizing naturally generated radiation of the target (scene) in the 3–14 μm wavelength as the source of received energy for detection or imaging. Microwave radars, on the other hand, operate in the active mode, transmitting energy from their own source toward the target. Detection occurs upon the reception of the reflected energy from the target. It should also be noted that there are passive microwave remote sensing systems, usually referred to as *radiometers*. They are used to passively detect electromagnetic energy of terrestrial and extraterrestrial origin. The antennas of these radiometers receive radiation emitted by the scene or object and reflected from the surroundings, such as the atmosphere. These passive microwave devices will not be discussed further in this book.

Another important difference between microwave and *infrared* (IR) electro-optical systems lies in the manner by which target or scene energy is collected for detection. Microwave radars transmit energy to the antenna, which in turn concentrates this energy into a narrow beam. This narrow antenna beam searches the target area either mechanically or electronically for the specific target. Mechanically scanned antennas, for example in a TV or raster scan, move the antenna beam by the physical motion of the antenna in the azimuth or the elevation planes. Electronically scanned antennas move the antenna beam electronically, with the radiating elements staying stationary while the antenna beam searches the target volume. The instantaneous position of the antenna beam in these radars is determined by the computer-generated phase of the transmitted voltage of each radiating element.

Figure 1.31 shows a typical antenna scan volume in the search and tracking mode of operation. This figure shows a long-range, as well as a shorter-range, scan pattern. The circles of the figure represent antenna beam positions. Two antenna beam coverage patterns are shown in this figure, in fact: a fast moving beam for short-range targets and a slower moving beam for longer-range targets. The right side of the figure also shows two individual target tracking antenna beams.

In electro-optical systems, the entire field of view of the optical lens is imaged in the focal plane of the system as shown in Figure 1.32. Resolution cell points of the object space, such as points A, B, and C, are imaged as shown. In present-day thermal imaging systems, because of the limited number of detectors, scanning mechanisms are used to image the object space through the receiving detectors. Thus, effectively, the detectors are time shared by the resolution elements of the object space. Figure 1.33 shows four points in the object space that are imaged through the scanning mechanism into a single detector. In electro-optical thermal imaging systems, the object space is usually imaged through the scanning mechanism into the detectors at a rate of 20 or 30 times/s.

Figure 1.34 shows the geometry of a line scanner onboard an aircraft. The scene directly below the aircraft is scanned through the rotating mirror of Figure 1.35 into the detector. In this manner, the entire ground strip below the aircraft is scanned line-by-line into the detector, which then will produce the display image. A great variety of one- and two-dimensional scanners are available in the electro-optical system applications as discussed in [10], [11], and [12]. Electro-optical systems are also used by satellites to produce thematic maps of the ground using wavelengths of 0.5–12 μm.

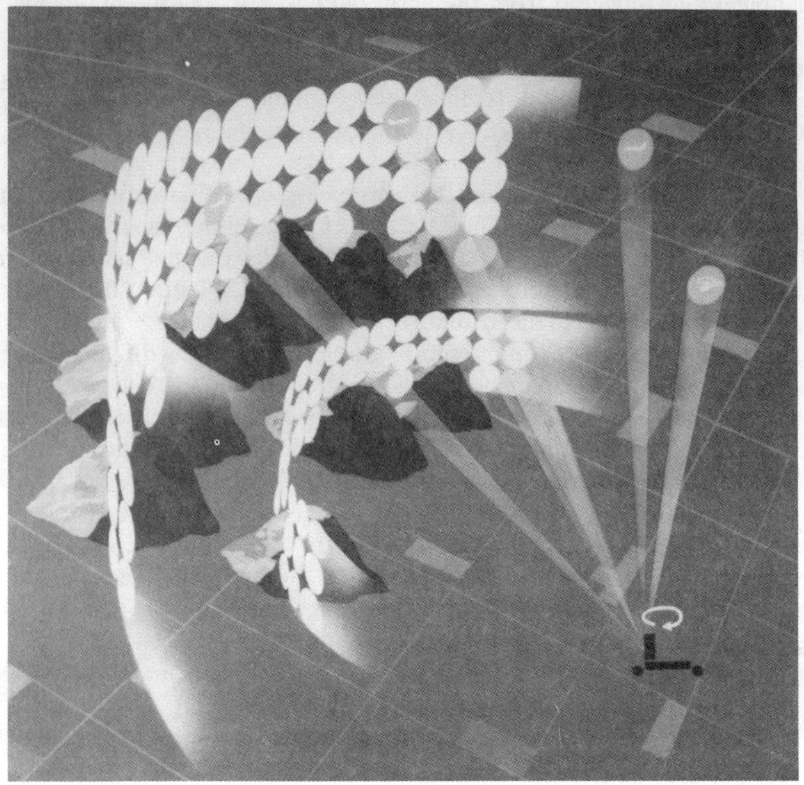

Figure 1.31 Antenna beam pattern coverage of a microwave radar in the search and tracking modes of operation.

We anticipate that future electro-optical thermal imaging systems will utilize *focal plane arrays* (FPA), where the number of detectors is increased substantially and an entire scene can be registered simultaneously. Figure 1.36 shows a conceptual sketch of an FPA system, where resolution cells (pixels) of the object space are directly imaged into a mosaic of detectors and subsequently displayed. Focal plane arrays are very similar to optical photography or video recording, where the entire object space is registered instantaneously into a photographic film or a solid-state imaging device, respectively. Solid-state high resolution imagers in the visible region of the frequency spectrum are called *charged coupled devices* (CCD). Reference [12] describes a 1320 × 1035 element CCD full frame imager with an optically active area that measures only 9 × 7 mm.

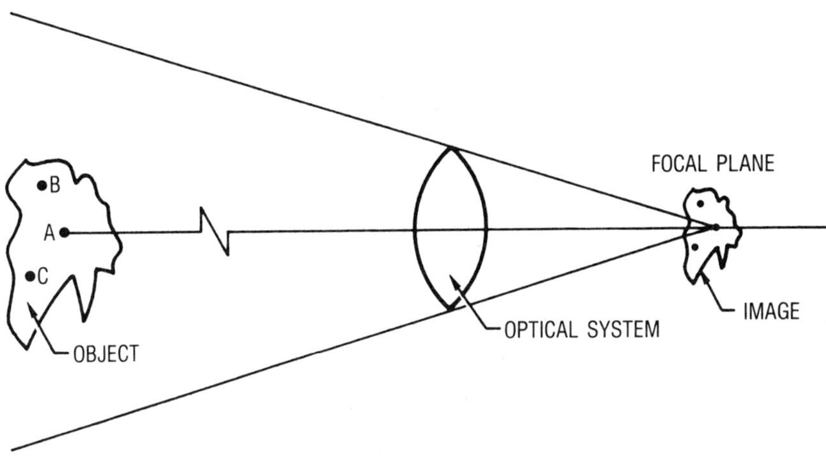

Figure 1.32 Object and image planes in an electro-optical system.

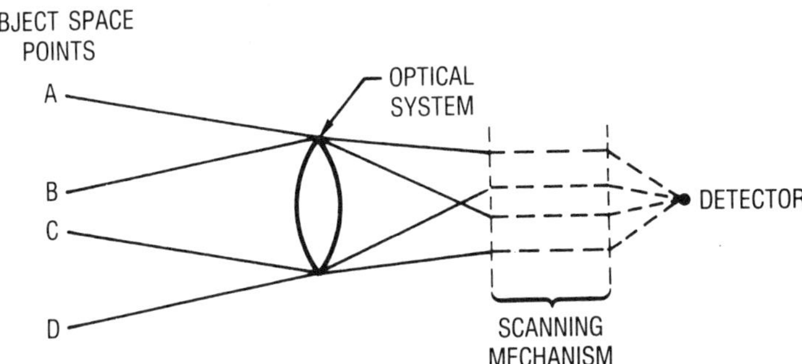

Figure 1.33 Object space resolution cells mapped into a single detector through the scanning mechanism.

According to [13], the Northrop Corporation is developing an FPA of 128 × 128 elements in the 3–5 μm wavelength region. The total of 16,384 detectors are read at a rate of 960 times/s. Before the display, 32 consecutive readings are averaged, at a rate of 30 frames/s. The detector elements are set at 0.0012 in. apart from each other. This set of detector arrays can be used as an imaging device at close ranges and as a "hot spot" detector at longer ranges.

In the next section we will discuss the general characteristics of sensor systems both from a performance and mechanization point of view.

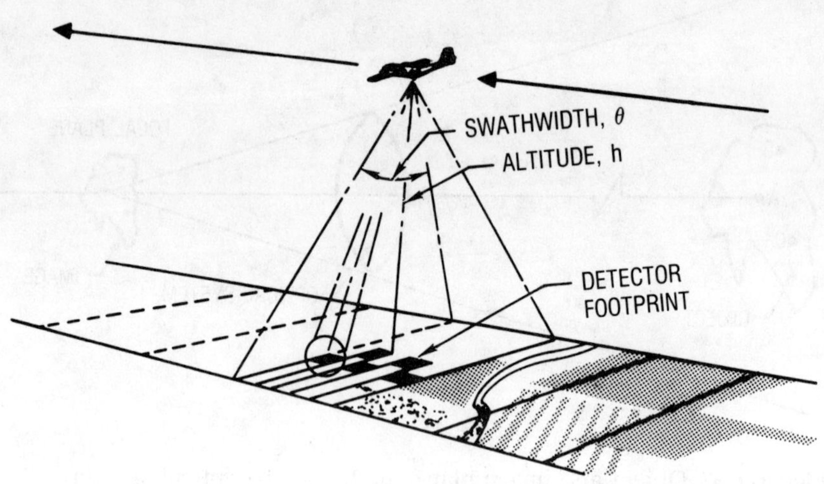

Figure 1.34 Line scanner of an electro-optical system producing an image of the ground below (adapted from [11]).

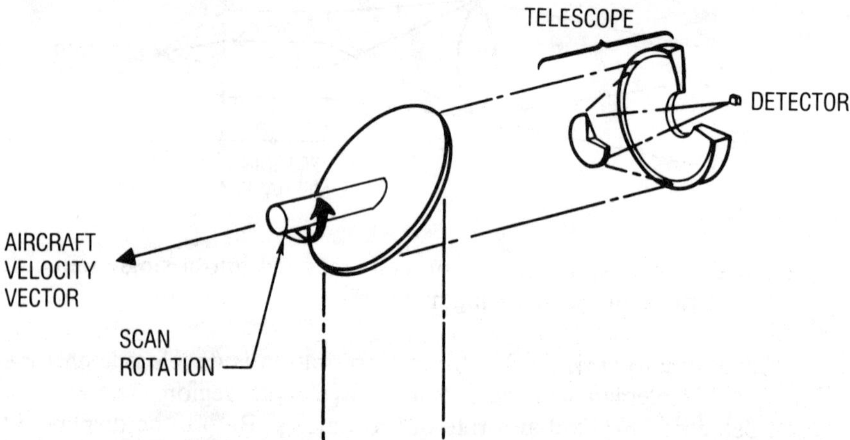

Figure 1.35 Rotary mirror scanner producing image of the ground below.

1.11 DISCUSSION

The overall sensor characteristics for the four sensor systems examined in this chapter are summarized in Table 1.6. Note that since most of the

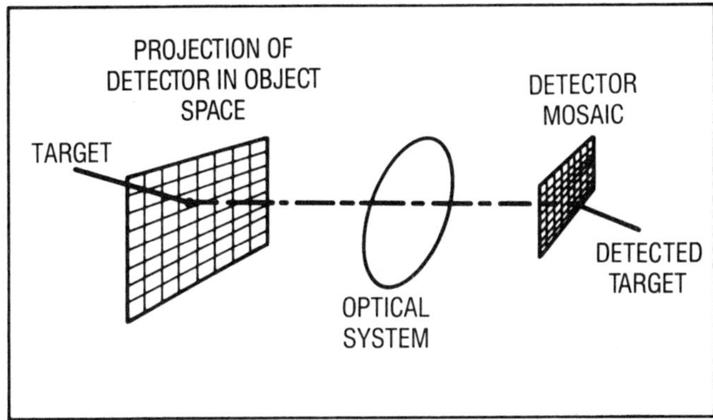

Figure 1.36 Conceptual presentation of the basics of focal plane array imaging.

sensors are designed for ground operation, atmospheric conditions are an important consideration in their design. The performance of spaceborne sensor systems is not usually limited by atmospheric conditions.

In recent years, target recognition capabilities have become an important consideration in the application of sensor systems. Presently, this capability is available only in the passive thermal imager and active laser radar systems primarily because of their higher angular resolution capabilities. This high resolution capability, the result of small beam widths, limits the ability of these systems to search large volumes for target detection.

In a number of cases, microwave or MMW radar systems can be used for initial search and acquisition of targets. This can be followed by laser radar tracking, resulting in extremely accurate angular positioning.

Table 1.7 outlines current applications of these sensor systems. EO systems have the unique capability of target recognition as compared to microwave and MMW sensor systems. The technology of these systems is growing rapidly, as work continues on focal plane arrays and high-power laser radar systems.

The sensors described in this book can be used individually or in combination, as in multiple sensor applications. Sensors used in combination usually complement each other in obtaining additional target information or improving information available from a single sensor, as discussed in detail in Chapters 7 and 8. For example, thermal imaging systems may be used in conjunction with laser radar range finders in tactical applications. In this case the thermal imaging provides a display of the target and background much like a TV picture. The operator can

Table 1.6 Overall Sensor Characterization

Sensor	*Active/ Passive*	*Detection Range (in atmosphere)*	*Atmospheric Effects*	*Range/ Range-Rate Values*	*Angular Position Resolution*	*Target Recognition Capability*	*Physical Charac-teristics*
Microwave	Active	Unlimited	Small	Available	Degrees or less	None (Except for SAR)	Large
MMW	Active	10 km	Some degradation	Available	Tenth of degree or less	None (presently)	Small
Thermal imager	Passive	10–15 km	Degraded	Not available	μr	Available	Small
Laser radar	Passive	5 km	Degraded	Available	μr	Available	Small

Table 1.7 Current Applications

Sensor	*Applications*	*Technology*
Microwave radars	Instrumentation, target surveillance and tracking, fire control systems, imaging systems, air traffic control	Mature
MMW radars	Short-range detection and tracking of stationary targets, gun pointing and fire control systems, target recognition	Emerging
Electro-optical systems	Target detection and recognition, short-range fire control systems, passive operation, multispectral ground mapping, high-power lasers	Rapid growth

then select the target of interest from this display and obtain target range by pointing the laser range finder in the direction of the target. Note that, being a passive device, the thermal imaging system does not obtain target range.

REFERENCES

1. S. A. Hovanessian, *Radar System Design and Analysis,* Artech House, Dedham, MA, 1984.
2. S. A. Hovanessian, "Detection of Non-Moving Targets by Airborne MMW Radars," *Microwave Journal,* March 1986.
3. Dean L. Mensa, *High Resolution Radar Imaging,* Artech House, Norwood, MA, 1981.
4. S. A. Hovanessian, "Electro-Optical System Design and Performance Equations," Hughes Aircraft Company Report No. TP-86-73-517, September 1986.
5. J. J. May and M. E. Van Zee, "Electro-Optical and Infrared Sensors," *Microwave J,* September 1983.
6. J. M. Lloyd, *Thermal Imaging Systems,* Plenum Press, New York, 1975.
7. K. Seyrafi, *Electro-Optical Systems Analysis,* Electro-Optical Research Company, Los Angeles, CA, 1973.
8. C. Bachmann, *Laser Radar Systems and Techniques,* Artech House, Dedham, MA, 1980.

9. A. V. Jelalian, "Laser Radar Theory and Technology," in *Radar Technology,* E. Brookner (ed.), Artech House, Dedham, MA, 1977, p. 339.
10. K. J. Button and J. C. Wiltse, eds., *Infrared and MMW,* Vol. 4, Academic Press, New York, 1981, p. 348.
11. C. T. Due, L. M. Peterson, "Optical-Mechanical Active/passive Imaging Systems," Environmental Research Institute of Michigan, May 1982.
12. ———, "KAF-1400, 1320 x 1035—Element Full-Frame CCD Imager," Microelectronics Technology Division, Eastman Kodak Company, Rochester, New York, 1987.
13. ———, "Northrop Developing Tactical Infrared Focal Plane Array," Aviation Week and Space Technology, November 2, 1987, page 82.

Chapter 2
Microwave Radar Systems

This chapter provides the basic principles and technology involved in the design and development of modern radar systems. Various types of radars are discussed, including low, medium, and high *pulse repetition frequency* (PRF) together with their governing principles, and several numerical examples are given. During the last few years, important advances have been made in digital radar signal processing methods. Applications of these methods are presented in this chapter. The discussion of radar and signal processing principles is followed by the derivation of the radar S/N ratio equation and its application in the design of radar systems. Another area of current radar interest is electronic scan antennas, which are discussed in this chapter. The final section includes methods of calculating the probability of detection of sinusoidal signal in noise and computing the threshold settings and false alarm rates.

2.1 INTRODUCTION

Pulse radars for measuring radar-target range have been in operation for more than 40 years. These radars measure range (distance between the radar and the target) by transmitting a pulse and measuring the time between the pulse and its return from the target. The elapsed time between transmitted and received pulses is proportional to the radar-target range. These radars, also known as low PRF radars, have a *pulse repetition frequency* (PRF) of anywhere between 30 and 4000 pulses/s.

Doppler radars, on the other hand, use the doppler (frequency) principle to measure radar-target range rate, rather than radar-target range as was the case in the low PRF radars. The range-rate is proportional to the frequency shift between the transmitted and the received signals.

The radar-target range in doppler radars is measured using specialized techniques, such as frequency modulation of the carrier frequency or multiple pulse repetition frequencies. Also known as high PRF radars, doppler radars have a characteristically high PRF, usually on the order of 65,000–300,000 pulses/s. Doppler radars are more complex to mechanize, but they present good discrimination between moving and nonmoving (e.g., ground) target returns. These radars are capable of detecting targets that are masked by ground clutter.

In recent years, the desirable features of low PRF pulsed radars (accurate range) and high PRF pulsed doppler radars (moving target indication) are combined in medium PRF radars. These radars use time to calculate radar-target range and doppler (frequency) information to indicate a moving target or calculate radar-target range rate. Medium PRF radars use PRFs in the range of 10,000–30,000 pulses/s.

In this chapter, we shall discuss the three types of radars: the low, high, and medium PRF just mentioned. Then, signal processing methods are discussed including analog signal processing methods as well as newer digital signal processing procedures. The radar equation and related S/N ratio calculations for target detection and target tracking also are presented. A numerical example is given with a supporting discussion of a target search and target tracking radar design. The final sections of the chapter provide a discussion of electronic scan antennas and radar systems and computation of the probability of target detection as a function of threshold setting and S/N ratio. Reference [1] is a general source of material in this chapter.

2.2 PULSED RADARS

Pulsed radars measure range (distance between the radar and the target) by transmitting a pulse and timing the returned pulse from the target as shown in Figure 2.1. Since electromagnetic energy travels at the speed of light, c:

$$t = \frac{2R}{c} \quad \text{or} \quad R = \frac{ct}{2} \tag{2.1}$$

where R is the radar-target range and t is the time between transmit and target return signals. The angular target information (azimuth and elevation of the target with respect to the radar) is obtained from the antenna position data at the time of target detection. With the values of these angles and range R, it is possible to establish the position of the target with respect to the radar.

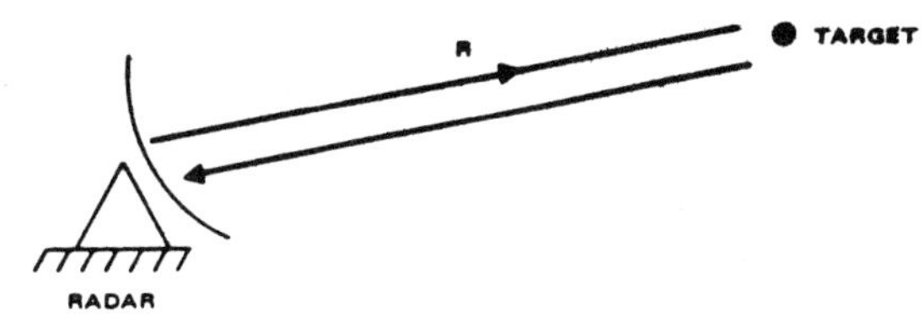

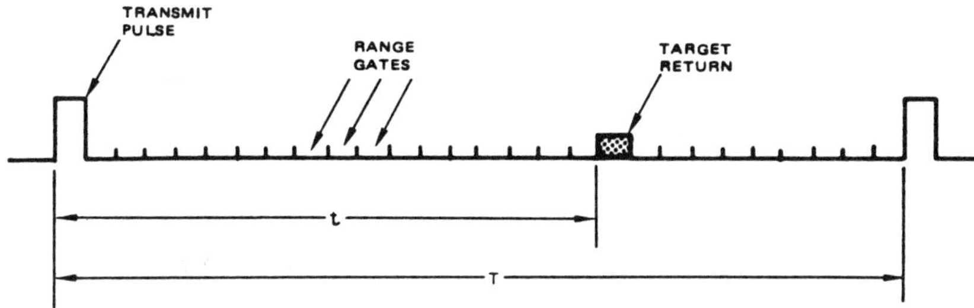

Figure 2.1 Pulse radar range measurement.

In the pulse radar of Figure 2.1, the maximum measurable range $R_{\max}$ is a function of T, the interpulse period, and is given by

$$R_{\max} = \frac{cT}{2} \tag{2.2}$$

Beyond this range, the elapsed time between transmitted pulse and target return will include multiples of interpulse period T. For this reason, we cannot determine correctly the value of elapsed time between transmitted and received pulses, making the measured range ambiguous.

Note that the transmitted pulse of Figure 2.1 is generated by pulsing a pure sine wave, as illustrated in Figure 2.2. The frequency of the sine wave f_o is usually in the 1000s MHz; for example, at X-band it can range between 8000 and 12,000 MHz (8×10^9 to 12×10^9 Hz).

The elapsed time between transmit and receive pulses, t, is measured by placing "range gates" all along the receive time as shown in Figure 2.1. The consecutive range gates open for the duration of a single pulse starting with each transmitted pulse. The presence of a signal in these gates corresponds to the elapsed time, t, used in Eq. (2.1) to compute the radar-target range. *In radars using a single antenna for transmitting and receiving, the receiver is turned off when the transmitter is on*; that is, *target returned pulses cannot be accepted when the transmitter is on.*

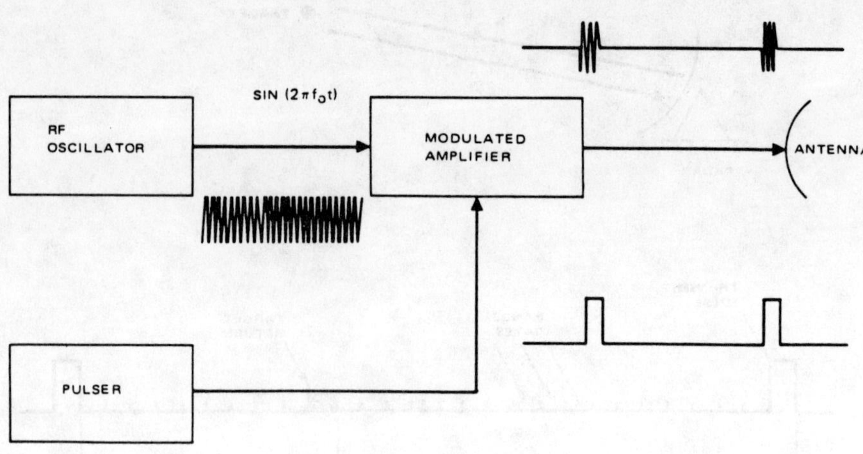

Figure 2.2 Generation of transmit pulses from a sine wave.

2.2.1 Example: Pulsed Radar

Consider a pulsed radar with a pulse repetition frequency of 600 pulses/s. Calculate (1) the radar-target range for $t = 320\ \mu s$ and (2) the maximum unambiguous range, R_{max}.

From Eq. (2.1), we calculate the radar-target range, R:

$$R = \frac{3 \times 10^5 \times 320 \times 10^{-6}}{2} = 48 \text{ km}$$

where 3×10^5 is the velocity of light in km/s. Using Eq. (2.2), we have

$$R_{max} = \frac{3 \times 10^5 \times (1/600)}{2} = 250 \text{ km}$$

2.2.2 Discussion: Pulsed Radars

In airborne applications, pulsed radars encounter difficulty in detecting targets under "lookdown" conditions. Figure 2.3 presents the encounter geometry and the detection process under these conditions. Figure 2.3a shows the radar-target geometry, while Figures 2.3b and 2.3c show the return pulse measurement and the range-azimuth display, respectively. From Figure 2.3b, note that ground returns emanate from ranges equaling

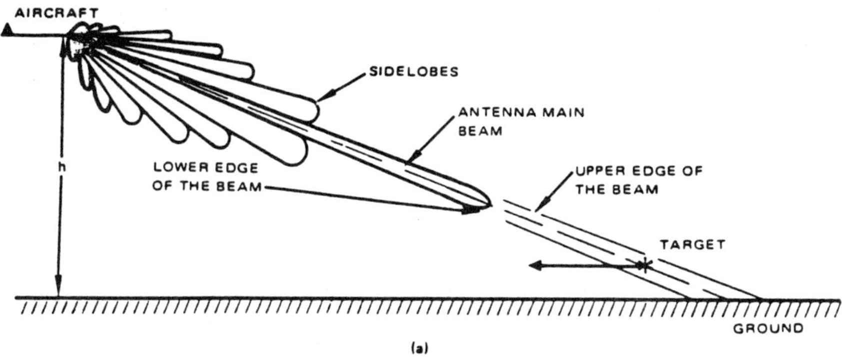

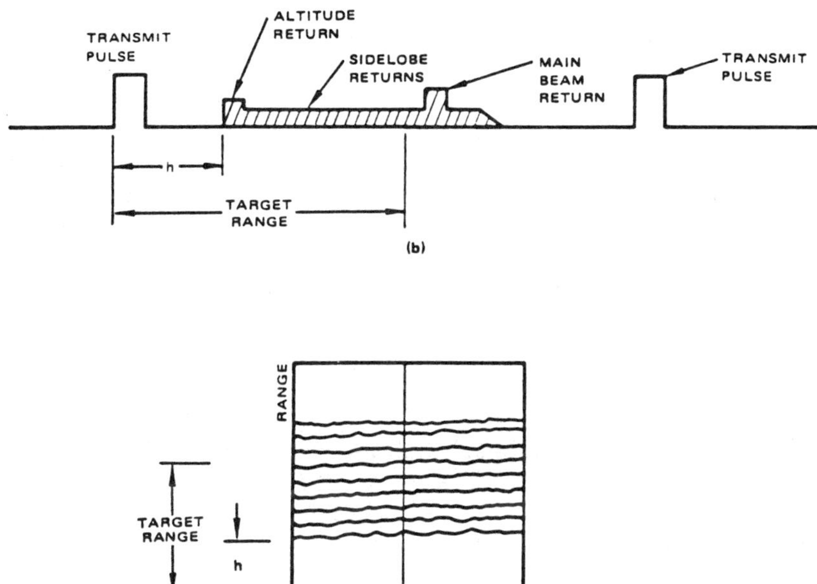

Figure 2.3 Pulse radar detection of ground targets.

aircraft altitude to ranges exceeding the slant range to the ground along the main beam. Most of these returns are due to antenna sidelobes.

Nevertheless, they interfere with the detection process, making it difficult to detect targets under lookdown conditions. The radar display characteristics are shown in Figure 2.3c. This figure depicts a cluttered display due to ground returns. The needs for reliable lookdown detection and increased target range led radar engineers to design doppler radars.

2.3 PULSED DOPPLER RADARS

To increase the maximum detection range of a pulse radar, we need to increase both the transmitted power per pulse and the interpulse period that determines the unambiguous radar-target range. Although the interpulse period can be increased, there are hardware limitations to increasing the transmitted power, especially in airborne radars under severe weight and volume constraints.

To increase the amount of transmitted energy, and thereby the detection range, as well as to detect moving targets in the presence of ground clutter, pulsed doppler radars were invented. These radars operate on the principle first observed by Christian Doppler early in the nineteenth century. Figure 2.4 illustrates this principle, which operates primarily on the frequency change caused by motion. Figure 2.4a shows the doppler effect on a nonmoving target, which results in *no* frequency change between transmit and receive waveforms. Figure 2.4b shows a closing target ($\dot{R} < 0$), which results in an increased frequency as compared to the transmitted frequency. Figure 2.4c illustrates the conditions for opening targets ($\dot{R} > 0$), where a decrease of frequency will be observed.

Using the doppler effect rather than timing the transmit-receive pulse allows us to use a very high PRF to increase the transmitted energy, as shown in Figure 2.5. This high PRF, however, eliminates the capability of the radar to obtain a radar-target range by timing transmit and target return pulses. High PRF radars use the doppler principle, which shows range rate, $\dot{R}$, as a function of transmit-receive frequency shift, given by the equation

$$\Delta f = \frac{-2\dot{R}}{\lambda} \tag{2.3}$$

where Δf is the difference between transmitted frequency and the target return frequency, and where λ is the wavelength. The negative sign of Eq. (2.3) is used to show that approaching targets (decreasing range or negative range rate, $\dot{R}$) produce positive frequency shifts Δf. Doppler radars are mechanized to measure $\dot{R}$ from the measured frequency shift

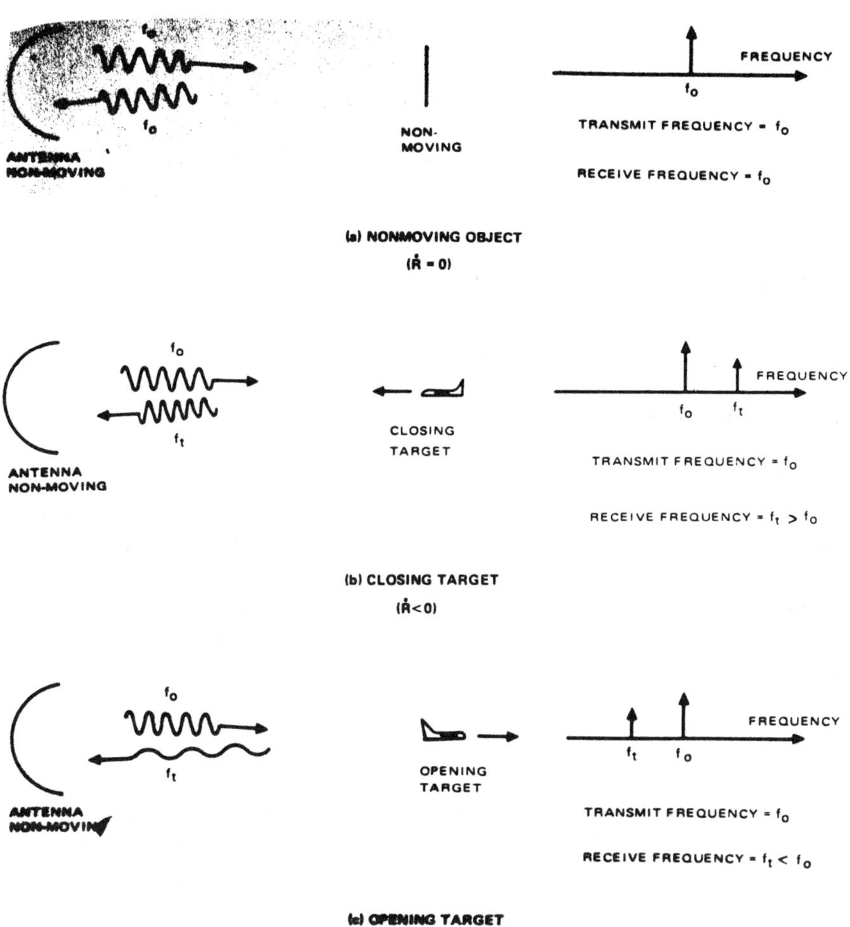

Figure 2.4. Doppler effect on targets.

Δf. Special techniques are used in these radars to measure radar target range, R.

2.3.1 Range Measurement

The increased number of pulses in high PRF radars destroys the ability of the radar to determine radar-target range by simple time discrimination ranging. To obtain radar-target range in high PRF radars, we may apply either frequency-modulation of the carrier frequency or multiple PRF

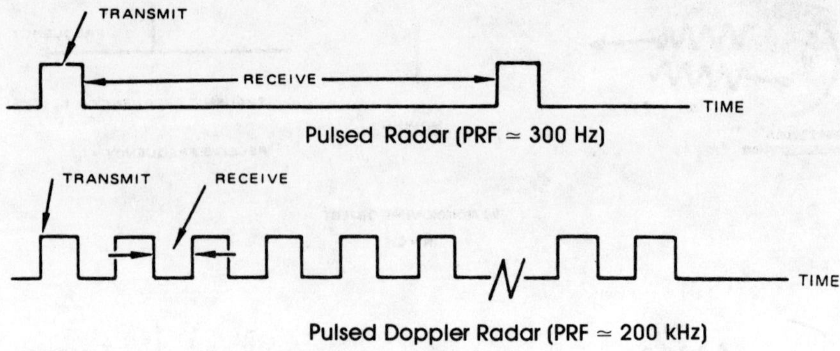

Figure 2.5 Pulsed and pulsed doppler radars.

ranging methods. Frequency modulation ranging methods are based on frequency shifts of the returned signal with respect to the transmitted signals. This frequency shift is translated into a time delay that corresponds to the radar-target range. Multiple PRF ranging methods, on the other hand, use time as the basis of measurements, like low PRF radars. The operating principles of the two ranging schemes are described in the following subsections.

2.3.1.1 Frequency-Modulation Ranging

With this method, the transmitted frequency is increased and decreased periodically. The frequency difference between the transmitted and received signals contains information from which the values of range, R, and range rate $\dot{R}$, can be derived. Figure 2.6 gives the relationship of the transmitted and received signal in a *linear* frequency modulation ranging scheme. Note that the received signal is delayed because of radar-target-radar round trip distance and frequency-shifted due to the radar-target closing rate. The applicable equations are as follows:

$$f_{b+} = \frac{2R\dot{f}}{c} + \frac{2\dot{R}}{\lambda}$$

$$f_{b-} = \frac{2R\dot{f}}{c} - \frac{2\dot{R}}{\lambda} \qquad (2.4)$$

where $\dot{f}$ is the modulation slope and R is the radar-target range.

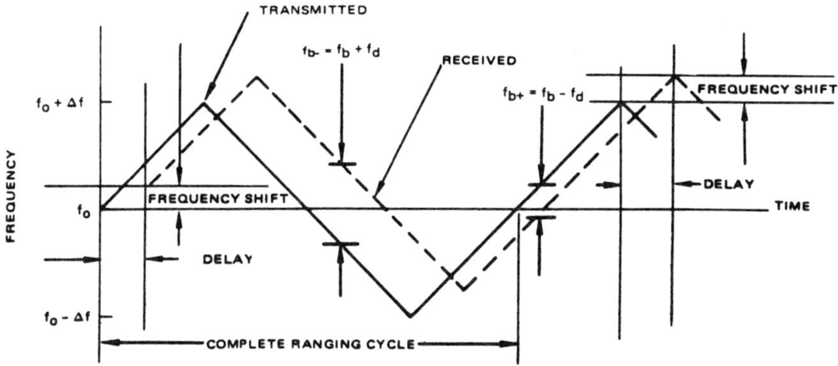

Figure 2.6 Linear modulation to obtain range and range rate values.

Adding the above equations, one obtains the expression for radar-target range, R:

$$R = \frac{c(f_{b+} + f_{b-})}{4\dot{f}} \tag{2.5}$$

Subtracting Eq. (2.4), one obtains range rate, $\dot{R}$

$$\dot{R} = -\frac{\lambda(f_{b-} - f_{b+})}{4} \tag{2.6}$$

The frequency differences f_{b-} and f_{b+} of Eqs. (2.5) and (2.6) can be measured as the difference between target return frequency and transmitted frequency. Note that one complete ranging cycle should occur during the target illumination period to solve Eqs. (2.5) and (2.6) for R and $\dot{R}$ values. The target illumination period, depending on the antenna sweep rate and the antenna beamwidth, is usually in the order of 10–50 ms.

Example

Consider an airborne radar using a transmit frequency of $f_o = 9000$ MHz with the frequency modulation ranging method of Figure 2.6. The complete ranging cycle is 20 ms and $\Delta f = 20$ kHz. For an f_{b-} of 6 kHz and an f_{b+} of 2 kHz, obtain the values of target radar range, R, and radar range, $\dot{R}$.

The slope, f, can be calculated as

$$f = \frac{\Delta f}{(T/4)} = \frac{20 \times 10^3}{(20 \times 10^{-3}/4)} = 4 \times 10^6 \text{ Hz/s}$$

where, from Figure 2.6, $T/4$ is the corresponding time period for Δf frequency change. The wavelength, λ, is equal to

$$\lambda = c/f_o = \frac{3 \times 10^{10}}{9 \times 10^9} = 3.333 \text{ cm}$$

Using Eqs. (2.5) and (2.6) together with these values, we obtain

$$R = \frac{3 \times 10^{10} (2 \times 10^3 + 6 \times 10^3)}{4 \times 4 \times 10^6} = 1.5 \times 10^7 \text{ cm}$$

$$= 150 \text{ km}$$

$$\dot{R} = -\frac{3.333 (6 \times 10^3 - 2 \times 10^3)}{4} = -3.333 \times 10^3 \text{ cm/s}$$

$$= -33.33 \text{ m/s}$$

2.3.1.2 Multiple PRF Ranging

In radars using time discrimination to calculate range, the pulse repetition frequency PRF determines the maximum unambiguous range that can be achieved by the radar. Figure 2.7 shows radar returns for two targets, widely separated in range, detected by a pulse radar. If target returns were not labeled, there would be no way of telling which transmitted pulse caused which target return. The method of multiple PRF ranging has been used successfully to overcome this difficulty. The principles involved in multiple PRF methods are illustrated in Figure 2.8. In this figure, the basic ranging frequency with an interpulse period of τ_r and the two PRFs of periods $\tau_r/3$ and $\tau_r/4$ are shown. Target returns of each PRF are illustrated by triangles. Note that, if timed from the beginning of each PRF transmission, only the return time, T, corresponding to the true target range will match for both PRFs.

In actual mechanization, during a single target illumination period, PRF 1 is transmitted first. The values of x in Figure 2.8, representing the time between the last transmitted pulse and target return, are obtained. Following this, PRF 2 is transmitted and the value of y is obtained. With the values of x and y, the radar-target range can be determined. Thus, in

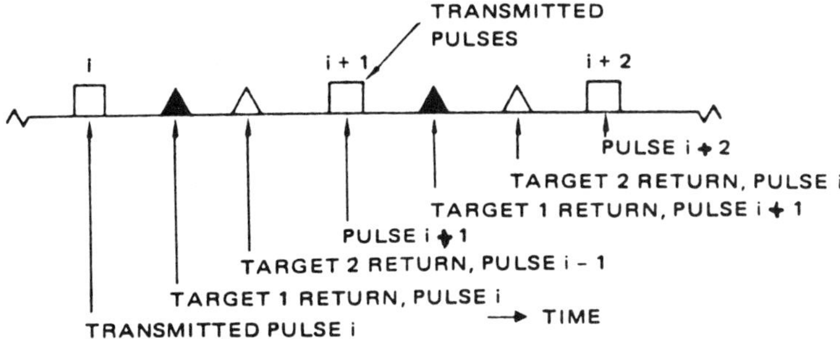

(▲ unambiguous, △ ambiguous)

Figure 2.7 Transmit and receiver pulses.

the multiple PRF mechanization, a great number of target return pulses are utilized in measurements of x and y, as compared to using the basic PRF where only a single returned pulse would be used.

Theoretically, a large number of PRFs can be used to approach the unambiguous range of the basic PRF of a multiple PRF ranging system. However, in practice, two or three PRFs are sufficient to resolve most ambiguous range problems. To construct an effective two-PRF ranging system [2], we may start by selecting a basic PRF, f_B, and an integer N (see Figure 2.8). The two ranging PRFs can then be calculated, for example, by

$$\begin{aligned} f_1 &= Nf_B \\ f_2 &= (N + 1)f_B \end{aligned} \tag{2.7}$$

Note that the multipliers N and $N + 1$ are relatively prime numbers, as they do not share a common factor.

Using the notation of Figure 2.8, we can write the expression for the time elapsed between transmission and the true target return. Designating this time by T, we can express it in terms of parameters of PRFs 1 and 2 as follows:

$$T = (n_1 + x)\tau_1 = (n_1 + x)\frac{1}{f_1} \tag{2.8a}$$

$$= (n_2 + y)\tau_2 = (n_2 + y)\frac{1}{f_2} \tag{2.8b}$$

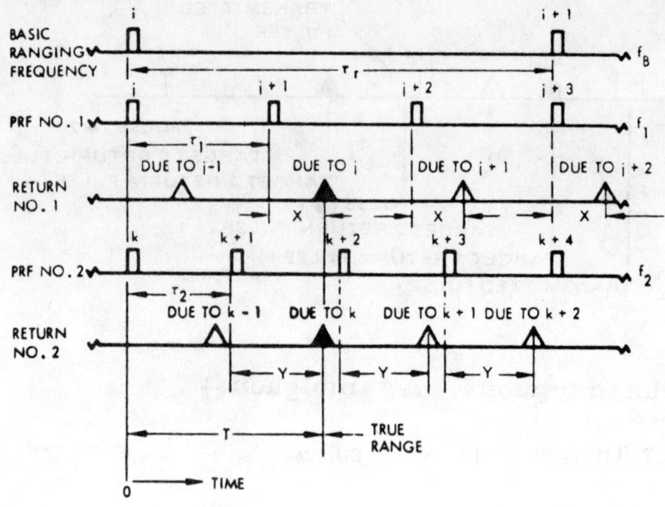

(▲ unambiguous, △ ambiguous)

Figure 2.8 Two-PRF ranging system.

where n_1 and n_2 are integers and represent the number of elapsed pulses between transmission and target return for PRFs 1 and 2, respectively; x and y are the corresponding fractions. Note that the radar system outputs only the values x and y. These values are usually in terms of range cell numbers (or range bins).

Given the PRFs f_1 and f_2 and the values of x and y from the radar, and using equations derived from Eq. (2.8), it is possible to calculate T of Eq. (2.8) and subsequently the radar-target range from $R = cT/2$. By using multiple PRF ranging, in effect we increase the unambiguous range from

$$R_{ua1} = \frac{c\tau_1}{2} = \frac{c}{2f_1}$$
$$R_{ua2} = \frac{c\tau_2}{2} = \frac{c}{2f_2} \tag{2.9}$$

corresponding to multiple PRFs f_1 and f_2 of Figure 2.8, to

$$R_{ua} = \frac{c\tau_r}{2} = \frac{c}{2f_B} \tag{2.10}$$

corresponding to the basic PRF, f_B. The advantage of multiple PRF ranging is the increased number of returned pulses and the resulting improvement in S/N ratio.

Additionally, the multiple PRF method of range calculation allows us to use a high PRF radar and calculate range in addition to range rate values, or moving target indication, which are available from doppler information.

Example

As an example of multiple PRF ranging, consider a radar with a maximum unambiguous range of $\approx$100 nmi. For $N = 79$, calculate the two PRFs, f_1 and f_2, the corresponding unambiguous ranges, and the values of x and y that will be obtained from the radar if the target is at a range of 53 nmi.

The basic ranging PRF, f_B, is calculated from

$$f_B = \frac{c}{2R} = \frac{161{,}875}{2 \times 100} \simeq 810 \text{ Hz}$$

where 161,875 is the velocity of light in *nautical miles* (nmi). The two PRFs are

$$f_1 = (79)(810) = 63.990 \text{ kHz}$$

$$f_2 = (80)(810) = 64.800 \text{ kHz}$$

The corresponding unambiguous ranges are

$$R_{ua1} = \frac{c}{2f_1} = \frac{161{,}875}{2 \times 63.990 \times 10^3} = 1.2648 \text{ nmi}$$

$$R_{ua2} = \frac{c}{2f_2} = \frac{161{,}875}{2 \times 64.800 \times 10^3} = 1.2490 \text{ nmi}$$

For a radar target range of 53 nmi, the elapsed time between the transmit and receive pulse will be

$$T = \frac{2 \times 53}{161{,}875} = 654.8 \times 10^{-6}\ \text{s}$$

Using this, we can calculate $(n_1 + x)$ and $(n_2 + y)$ from Eqs. (2.8) as follows:

$$(n_1 + x) = Tf_1 = 41.902$$

$$(n_2 + y) = Tf_2 = 42.432$$

The values of x and y will be the decimal fraction of these numbers, that is:

$$x = 0.902$$
$$y = 0.432$$

These values correspond to the number of the "range gates" where the target return pulse is timed divided by the total number of range gates. Note that by using multiple PRFs, the unambiguous range increased to 100 nmi from $\approx$1.25 nmi given earlier.

2.3.2 Detection in Clutter and Ground Returns

In addition to the higher number of transmitted pulses, which results in longer detection ranges, doppler radars obtain very accurate range rate, $\dot{R}$, values as their primary measurement. This characteristic is exploited in separating ground clutter (nonmoving targets) from actual moving target returns. This is a very desirable feature, especially in airborne radars, where ground clutter interferes with actual target detection (see Figure 2.3). Figure 2.9 shows an aircraft-target encounter geometry together with the corresponding amplitude-frequency diagram of the returned target signal using a doppler radar. Note from this figure that the relative velocity, with respect to the aircraft, of stationary objects (ground) are smaller than those of moving targets. Thus, all ground clutter return of Figure 2.9b occurs at velocities below those of the aircraft. In Figure 2.9b, the mainlobe clutter return, which results from the interception of the antenna mainbeam and the ground, is considerably larger than other returns. The altitude return at frequency f_o also accounts for considerable power return. In actual radar mechanizations, mainlobe clutter and altitude returns are usually taken out by placing filters in proper locations (see Figure 2.9b).

Since the doppler shift is proportional to the velocity, the ground return (clutter region) can be separated from the regions where target

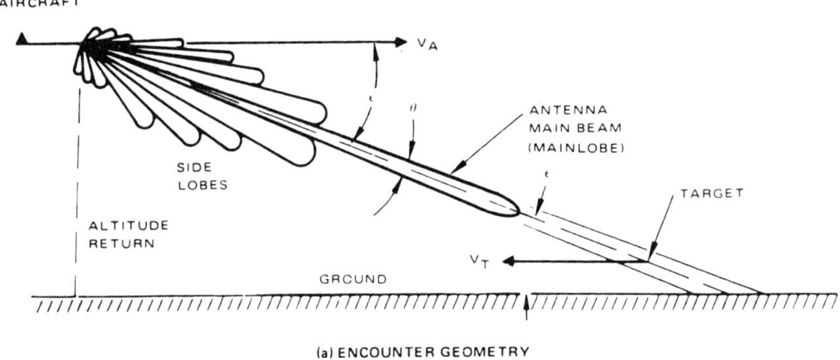

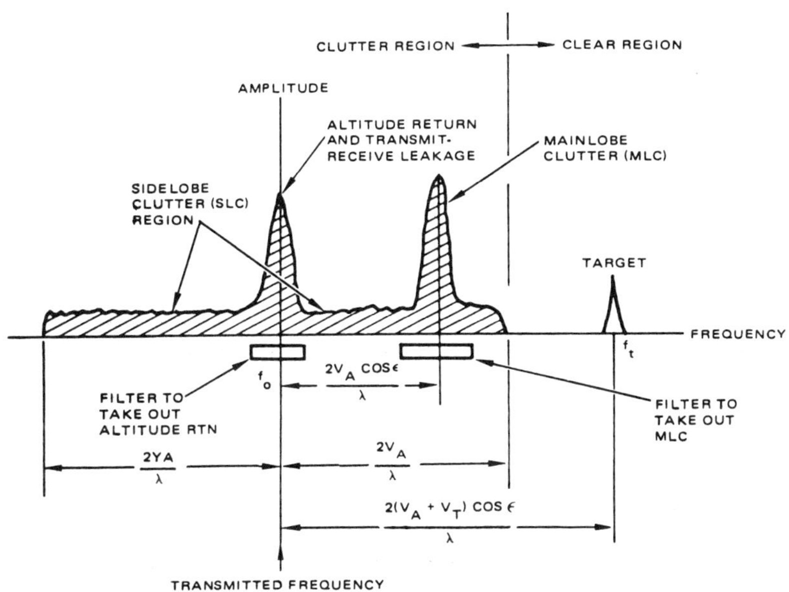

Figure 2.9 Amplitude-frequency characteristics of a doppler radar.

returns are expected. Figure 2.10 summarizes target detection capability of doppler radars under several encounter geometry conditions. These figures are shown for "lookdown" geometries where the radar mainbeam intercepts the ground. Under "lookup" conditions, where the radar beam

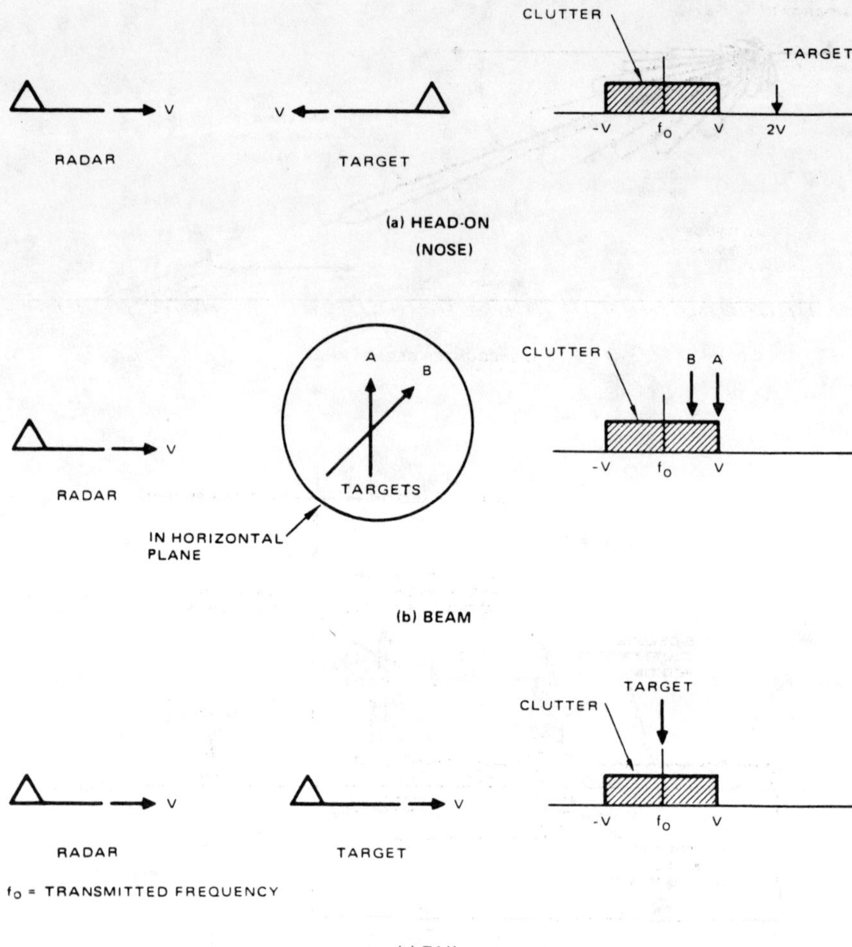

Figure 2.10 Encounter geometry and target return under clutter conditions for a doppler radar.

does not intercept the ground, ground clutter is smaller and targets can be detected more easily.

Under the lookdown clutter conditions of Figure 2.10, head-on closing targets (Figure 2.10a) can be detected easily, as their frequency shifts are beyond the frequency shift of the clutter region. The doppler radar, however, will have difficulty detecting beam targets of Figure 2.10b, as these targets are in the mainlobe clutter frequency region. The target return in this case will be competing with the stationary ground ahead of the aircraft. The tail geometries of Figure 2.10c will also produce difficult-to-detect targets under lookdown conditions. With a closing rate of zero ($\dot{R} = 0$), these targets will appear at the transmitted frequency f_o. The target return in this case is competing with the ground return directly below the aircraft (altitude return) and the transmitter noise.

Pulsed radars, on the other hand, will have difficulty detecting targets under all lookdown conditions, such as those illustrated in Figure 2.10. These radars, however, show a better performance for *coaltitude* beam and tail targets (Figures 2.10b and 2.10c) than the doppler radar, which looses the target completely in clutter. In addition, pulsed radars are used to measure range to ground and ground mapping. For these reasons, modern radars often are built with *both* pulsed and pulsed doppler capabilities.

Under tactical conditions, the doppler mode is used for long-range detection and tracking of incoming targets at all altitudes, while the pulse mode is used for the coaltitude or lookup beam and tail geometries similar to those of Figures 2.10b and 2.10c.

2.4 COMPARISON OF PULSED AND PULSED DOPPLER RADARS

The general background of pulsed and pulsed doppler radars, presented earlier, allows the comparison shown in Table 2.1. As developed in the preceding discussion, note that these radars exploit the doppler effect, resulting in range rate as the primary measured parameter. On the other hand, pulse radars use time discrimination between transmitted and received signals, resulting in range as the primary measured parameter. The ground clutter discrimination is accomplished more efficiently in high PRF doppler radars. The average transmitted power of high PRF radars is considerably higher than that of the corresponding low PRF radars under the same peak power restriction.

The advantages of high PRF doppler radars (as described earlier) are obtained at the expense of more complex mechanization than is required by pulsed radars. One area of difficulty is the frequency stability of the transmitted signal and the subsequent signal-processing mechanization to maintain the frequency shift achieved between transmit and receive frequencies. This signal processing should be accomplished during the frequency modulation cycle of Figure 2.6. Advances in RF transmitter power supply units, such as *traveling wave tubes* (TWTs), have made it possible to achieve frequency stability of the transmitted waveform during the target illumination period—a requirement in the mechanization of doppler radar systems.

Table 2.1 Comparison of High PRF Doppler Radar and Low PRF Pulsed Radar

Item	*High PRF Doppler Radar*	*Low PRF Pulsed Radar*
Average power	High	Low
Range rate measurement	Primary	Secondary
Range measurement	Secondary	Primary
Range determination	Sophisticated	Simple
General system characteristics	Complex	Simple
Detection in clutter	Excellent	Fair

Recently, there has been an attempt to combine the desirable features of the high PRF radars, which basically operate on frequency discrimination, and the low PRF radars, which operate on time discrimination. Medium PRF radars were developed to maximize the advantages of both systems. These radars operate in the ranges of 10–30 kHz PRFs and are mechanized to utilize both time and frequency information. To explain the workings of these radars, we need to describe time-frequency relationships of the transmitted and returned pulses.

2.5 FREQUENCY DOMAIN REPRESENTATIONS

The general frequency-time characteristics of various radars are shown in Figure 2.11. In the *continuous wave* (CW) radar of Figure 2.11a, the time-domain representation shows a continuous sine wave, sin $2\pi f_o t$, representing the transmitted waveform. In the frequency domain, the transmitted frequency is represented by a single carrier frequency, f_o. The coherent pulse radar time domain illustrated in Figure 2.11b shows a train of pulses that are cut from a continuous sine wave. Note that the phase relationship between pulses is maintained. The length of a given pulse is labeled t, and the period between consecutive pulses is labeled T. The reciprocal of this pulse period ($1/T$) is the PRF, f_P. The frequency domain representation of the pulse train of Figure 2.11b will consist of several distinct frequency lines at a "distance" f_P apart, as shown. Figure 2.11c is a diagram similar to 2.11b, but this time for a high PRF radar with a larger f_P. This will, of course, increase the "distance" between frequency lines, as shown in the corresponding frequency diagram.

Figure 2.11d describes a low PRF, *noncoherent* pulsed radar in time and frequency diagrams. Note that the phase relationship of the transmitted waveform is *not* maintained. Frequency representation, therefore,

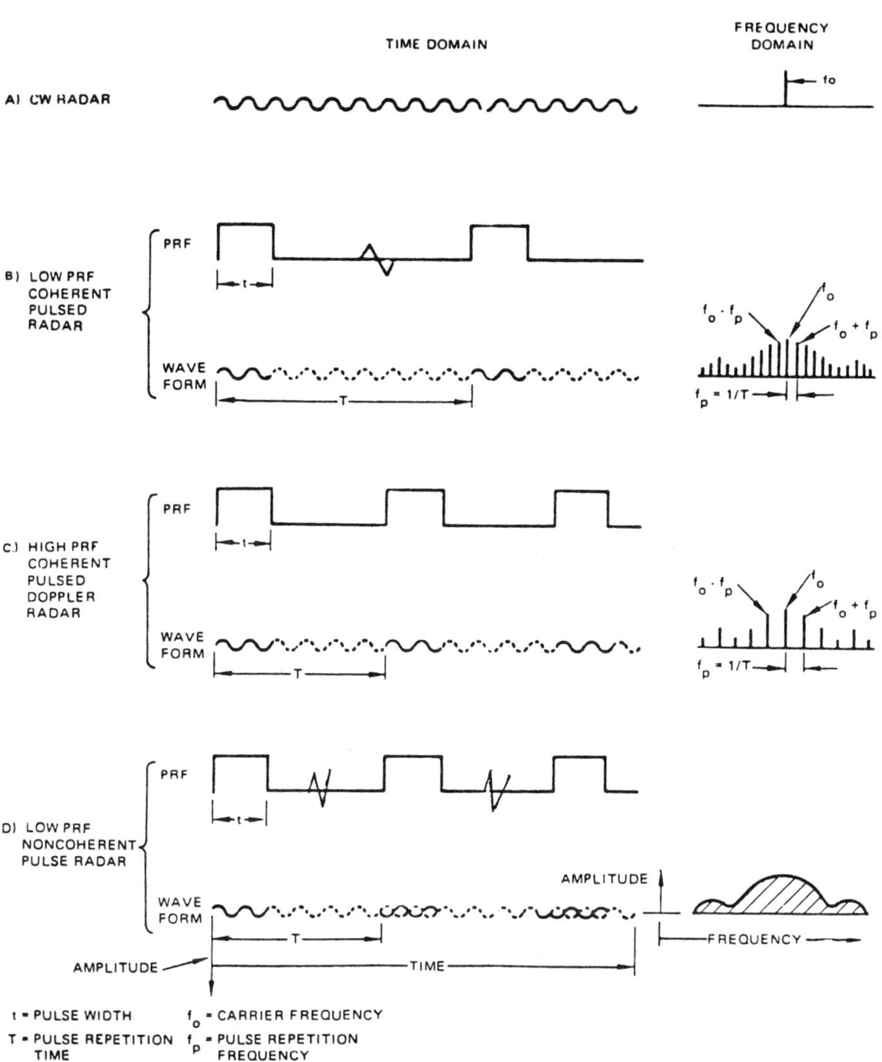

Figure 2.11 Types of radar transmissions.

is not in distinct lines as given in the coherent case of Figures 2.11b and 2.11c. The noncoherent waveform of Figure 2.11d represents, in effect, a "smearing" of frequencies over the frequency region rather than distinct frequency lines. Since high and medium PRF radars use frequency separation in the measurement of range or range rate, the coherency of the transmitted waveform is essential to their operation. In low PRF radars, where timing is used for radar-target range calculation, the returned target pulses can be integrated noncoherently (postdetection integration); therefore, coherency of the transmitted waveform is not essential to the operation of low-PRF radars.

Figures 2.12 and 2.13 show expanded versions of Figures 2.11b and 2.11c for a low PRF radar with a PRF of 100 Hz and a high PRF radar with a PRF of 300 kHz. Note that in the low PRF radar, target return signals over the entire frequency band are processed in the receiver. In the high PRF radar of Figure 2.13, only the target return due to the centerline frequency at f_o is processed, and the rest of the returns are filtered out.

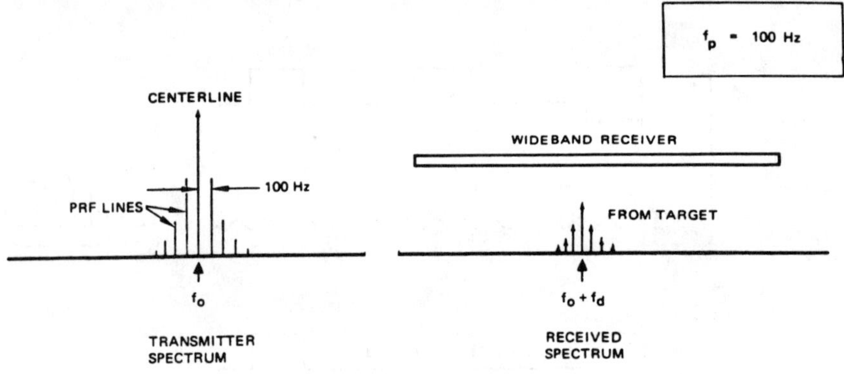

Figure 2.12 Transmitted and received spectrum of a coherent low PRF radar.

The frequency domain representation of Figure 2.13 under "look-down" conditions, shown with ground clutter present in Figure 2.14. This figure is drawn to point out the fact that all the PRF lines also receive their own clutter and target returns, although they are filtered out before entering the receiver. The approximate frequency difference of various returns at *X*-band is also shown in Figure 2.14.

2.6 MEDIUM PRF RADARS

Medium PRF radars are constructed utilizing both time discrimination for ranging and frequency discrimination for moving target indication or range rate calculation. Medium PRF radars usually are built in the PRF

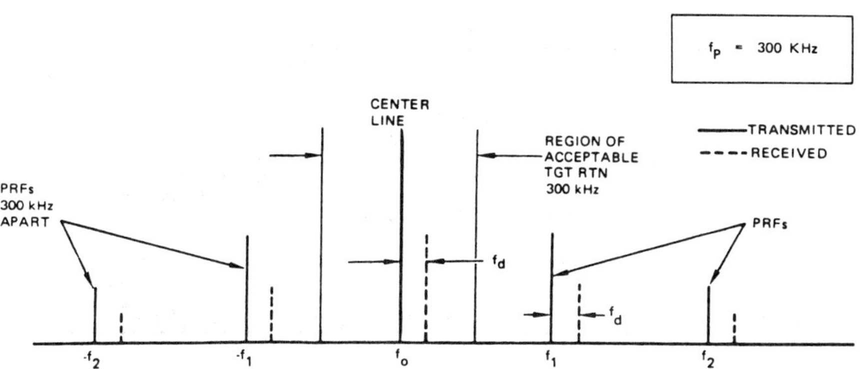

Figure 2.13 Transmitted and received spectrum of a high PRF radar.

range of 10–30 kHz [3]. In the following discussion, coherency of the transmitted signal is assumed.

Consider a medium PRF radar with a PRF of 15 kHz. This PRF is selected to put the *mainlobe clutter* (MLC) return of Figure 2.14 at the PRF line. The frequency spectrum of this medium PRF radar is shown in Figure 2.15 as is the target return, which is repeated for each PRF line. Note also that the target return competes with sidelobe clutter for detection when the radar is at low altitude and with system noise when the radar is at high altitude. Thus, in order to increase target detection capability in medium PRF radars, it is crucial to reduce sidelobe clutter levels. These levels can be effectively reduced by reducing antenna sidelobes.

Limiting the area of detection to the frequency range between the centerline and the first PRF line, we may place detection filters as shown in Figure 2.15. An examination of Figure 2.15 reveals frequency ambiguity much like the range ambiguity discussed earlier. We cannot be sure whether the target return shown in the "usable detection area" is due to the centerline or one, two, three, . . ., PRFs before the centerline. To resolve this frequency ambiguity, we may utilize PRF switching similar in principle to the previously discussed multiple-PRF ranging method.

Consider the case in which we use PRFs of 15 and 12 kHz. The frequency spectrum from these PRFs is illustrated in Figure 2.16. The location of detection filters is shown by dots. A target with a doppler shift of 22 kHz can be seen on the eleventh filter with the 15-kHz PRF and on the fourteenth filter with the 12-kHz PRF. Knowing the location of these filters and having the PRF values of 15 and 12, we can resolve

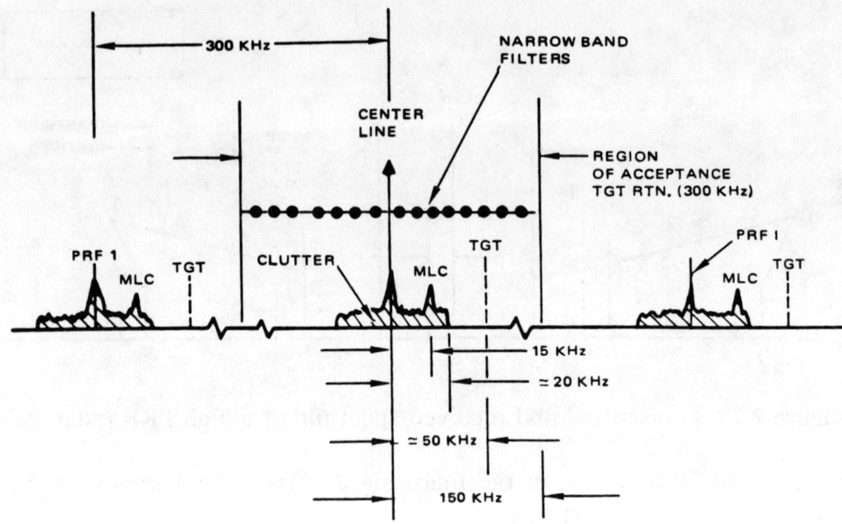

Figure 2.14 Transmitted and received spectrum of a high PRF pulsed doppler radar including clutter return.

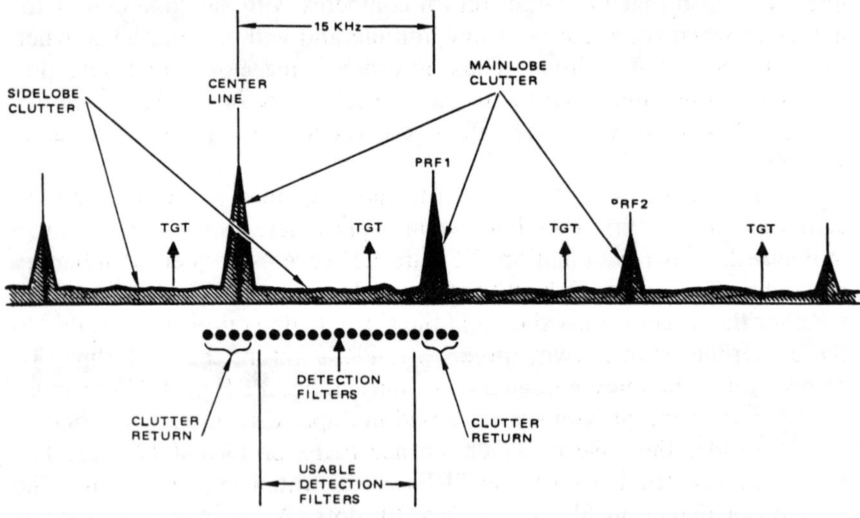

Figure 2.15 Medium PRF radar frequency spectrum in presence of ground clutter.

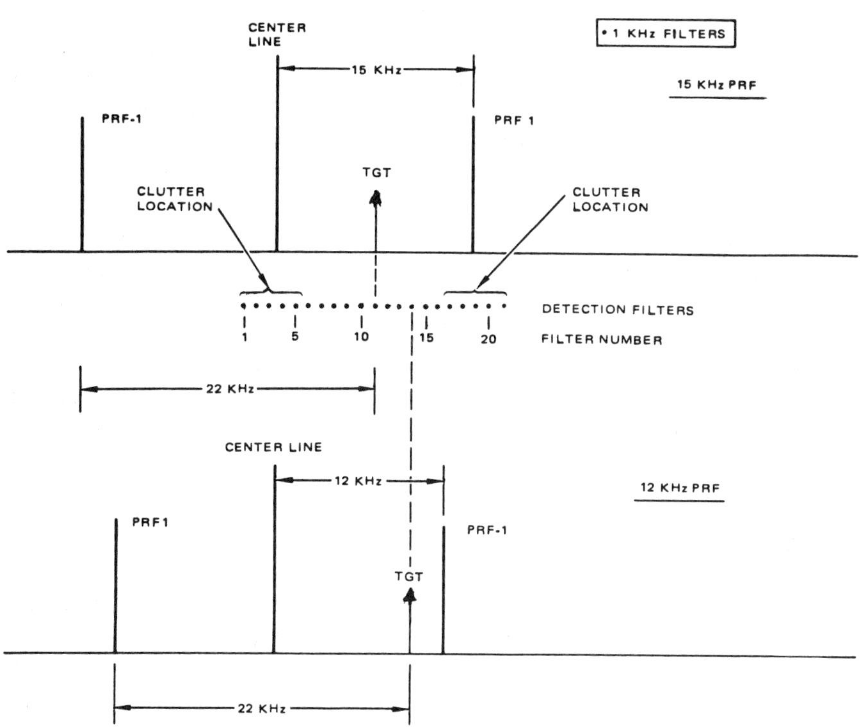

Figure 2.16 Multiple PRF target doppler calculation.

the frequency ambiguity and calculate radar-target range rate. The PRF selection to resolve doppler shift ambiguity is made on the basis of usable detection area. This selection also includes a consideration of the location of clutter and the fact that targets in the clutter filters will not be detected. It is sometimes necessary to go to many PRFs to ensure that all targets be detected, in the usable detection filters of Figure 2.15, in at least enough of these PRFs to permit resolution of true velocity.

In the medium PRF radars, the radar-target range is usually calculated by using the multiple PRF ranging methods described previously. The PRFs used for multiple ranging are usually not the same as those used for frequency ambiguity resolution, since the criteria of selection are different. In these radars, two sets of PRF may be used: one for range calculation and one for range rate calculation.

The selection of PRFs in a medium PRF radar can also be accomplished through a computer simulation of several sets of candidate PRFs. In this method, several sets of range and doppler PRFs are selected, and each set is tested by a complete simulation for clear range and velocity regime. Targets are detected in the clear regime and are not detected in the "blind zones." Figure 2.17 is a range-doppler plot showing typical simulation outputs. This figure shows "blind zones" as well as clear regions. Range blind zones occur when target return pulses are eclipsed by transmitted pulses; that is, target return pulses arrive when transmitter is on. Doppler blind zones occur when doppler frequency shifts of the target returns coincide with mainlobe clutter or altitude returns. This method can be used when independent measurements of range, R, and range rate, $\dot{R}$, are desired. Note that range rate $\dot{R}$ can also be obtained by differentiating range R; this $\dot{R}$ information can be used in the selection of doppler PRFs and the resolution of doppler ambiguity.

Example

As an example of a medium PRF radar, let us assume a basic frequency resolution PRF of 15 kHz. Assume that filters are placed every 1000 Hz and that the mainlobe clutter and altitude return take up a 3000-Hz region (three filters). To ensure that targets will not fall in the clutter filters, we may select two additional PRFs 3 kHz apart; that is, 12 and 18 kHz (see Figure 2.18). From this figure, we note that MLC is displaced 15 kHz from the center line and the PRF lines. Also note that a target with a doppler shift of 30 kHz will fall in the clutter filters of the 15-kHz PRF. This target, however, will be in the clear region of the 12 and 18 kHz PRFs. Note that the location of MLC is known by the data-processing computer, and the input from corresponding clutter filters can be ignored.

To calculate the doppler shift of the target, we proceed as follows. In the 12-kHz PRF, the target will fall in filter number 6; in the 18-kHz PRF, it will fall in filter number 12 (see Figure 2.18). To obtain the correct target doppler shift, we add the corresponding PRF increments to the filter numbers and seek a match between resulting values; that is:

PRF	*Doppler Shift*
12 kHz	$6 + 12 + 12 = 30$
18 kHz	$12 + 18 = 30$

Thus, the ambiguity is resolved; the correct target doppler shift is 30 kHz.

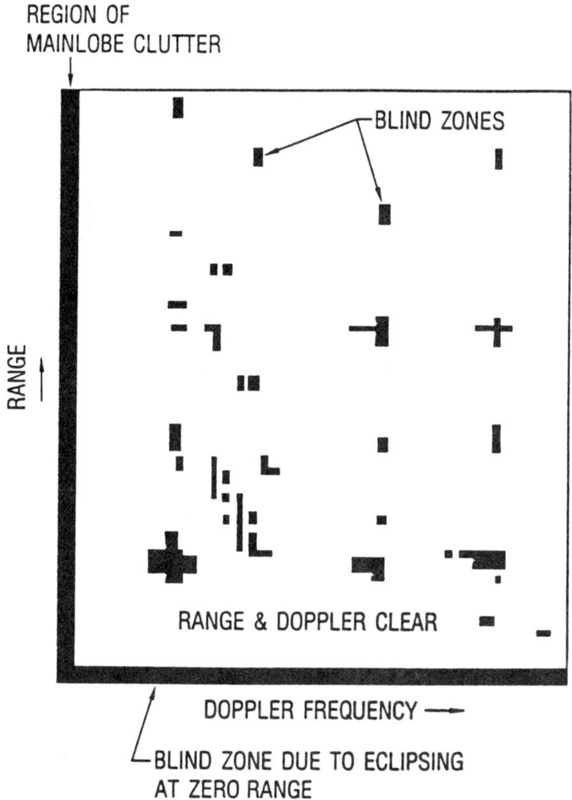

Figure 2.17 Selection of PRFs using clear range and doppler regions.

Ranging PRFs are selected by using the criteria described in Eq. (2.7). For example, taking the basic PRF (f_B = 810 Hz) for the 100-nmi unambiguous range and selecting $N = 17$, we obtain

$$f_1 = (17)(810) = 13.770 \text{ kHz}$$

$$f_2 = (18)(810) = 14.580 \text{ kHz}$$

Using these PRFs, assume that we have measured an $x = 0.508$ and a $y = 0.832$ [Eq. (2.8)] from the radar. We wish to calculate the corresponding radar-target range.

The solution can be obtained by matching the time (distance) between transmit and receive pulses from each of the two PRFs [Eq. (2.8)]. The

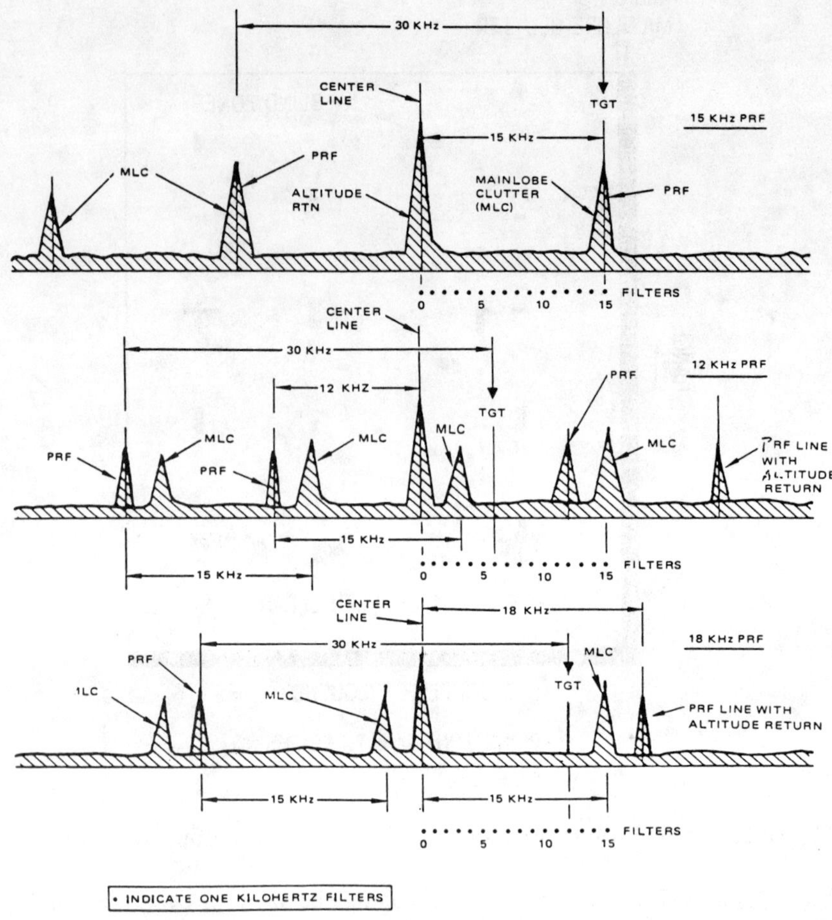

Figure 2.18 Frequency resolution PRFs of the medium PRF radar example.

unambiguous range (equivalent distance between two consecutive pulses) can be computed for each PRF as follows:

$$R_{ua1} = \frac{c}{2f_1} = \frac{3 \times 10^5}{2 \times 13770} = 10.893 \text{ km}$$

$$R_{ua2} = \frac{c}{2f_2} = \frac{3 \times 10^5}{2 \times 14580} = 10.288 \text{ km}$$

Using this and the values of x and y, we can write the radar-target range:

$$\text{Range} = (0.508)(10.893) + n(10.893) \qquad \text{PRF 1}$$
$$= (0.832)(10.288) + n'(10.288) \qquad \text{PRF 2}$$

when n and n' are integers. Computing range for values of n and n' = 1, 2, . . . , using two decimal places, we obtain

$$\text{PRF 1} \quad R = 16.43, 27.32, 38.21, 49.11, \mathit{60.00}, 70.89, 81.78, 92.68, \ldots$$
$$\text{PRF 2} \quad R = 18.88, 29.14, 39.42, 49.71, \mathit{60.00}, 70.29, 80.58, 90.86, \ldots$$

We note a match between range values calculated for PRFs 1 and 2 at 60 km; therefore, 60 km represents the measured radar-target range. Thus, two ranging PRFs, together with frequency resolution PRFs, will result in radar-target range and also moving target indication or radar-target range rate.

2.7 RADAR SIGNAL PROCESSING

From the discussion provided in this chapter, it is obvious that the radar signal processing involves (1) timing the elapsed time between transmit and receive pulses to obtain radar-target range, and (2) comparing the transmitted frequency, f_o, with the target return frequency, f_t, to obtain the resulting frequency shift, $\Delta f = f_t - f_o$ (doppler shift and range rate). This frequency shift can be obtained by analog methods through tuned filters or, more recently, by digital methods through the use of digital computers [4].

Figure 2.19 gives the general analog data-processing diagram of a pulse radar, and Figure 2.20 shows the signal processing of a doppler radar after the IF amplification stage of Figure 2.19. In Figure 2.20, each doppler filter is followed by a detector prior to target display indication.

The duplexer of Figure 2.19 switches the transmitter between transmit and receive functions. Note that when the transmitter is *on*, the receiver is *off*, and vice versa. The target return comes in at frequency $f_o + f_d$, where f_d is the frequency shift. Target return is then amplified and reduced in frequency from RF (1000s of MHz) to IF (30 MHz) by

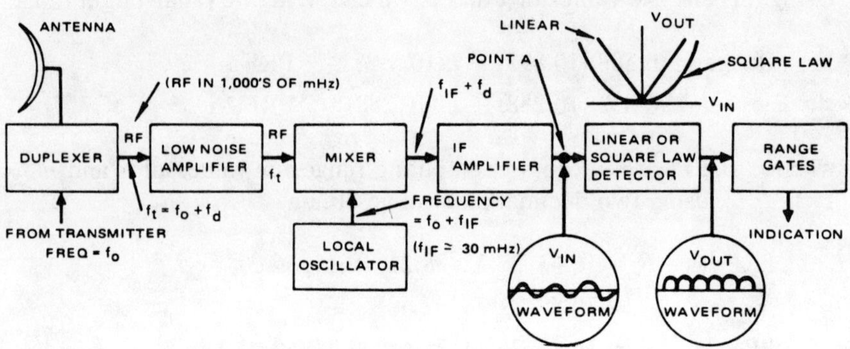

Figure 2.19 Superheterodyne radar receiver.

the mixer of Figure 2.19. This process is accomplished through multiplication (hence the name heterodyne) of incoming waveform $\sin 2\pi(f_o + f_d)t$ by the reference waveform of the local oscillator $\sin 2\pi(f_{IF} + f_d)t$ and retention of the low frequency term $(f_{IF} - f_d)$ of this product. Thus, the targets with positive doppler shifts appear at frequencies lower than the IF frequency rather than $(f_{IF} + f_d)$. This symmetrical inversion, however, can be easily tracked. This signal is amplified and, in pulse radars, passes through a detector prior to scope display. In doppler radars, the IF amplification stage is followed by a set of doppler filters for doppler shift detection, as shown in Figure 2.20 and discussed previously.

In radars that utilize both range gates and doppler filters to obtain radar-target range and range rate, respectively, signal-processing mechanization involves a combination of Figures 2.19 and 2.20. Figure 2.21 shows the signal-processing block diagram following the IF amplifier stage of Figure 2.19 for a pulse doppler radar, where both range and range rate values are obtained simultaneously from the returned signal. Note that, in this case, a set of doppler filters is provided at the output of each range gate.

In newer radars, Fourier transforms and digital signal processing methods replace the doppler filters for range-rate computations as discussed in the next section.

2.8 DIGITAL SIGNAL PROCESSING

As an example of the digital signal processing procedure, consider the target return signal, f_t, of Figure 2.19. To follow the procedure numerically,

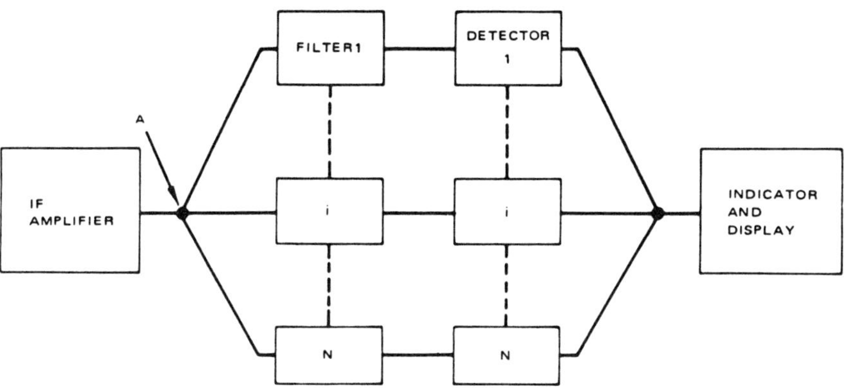

Figure 2.20 Arrangement of detection filters.

assume that f_t is the sum of the X-band transmitted frequency f_o = 10,000 MHz and a target doppler shift f_d of 6200 Hz. If we assume a unity amplitude, we can write the received signal as

$$\begin{aligned} V_r &= \sin [2\pi(f_o + f_d)t] \\ &= \sin [2\pi(10{,}000{,}006{,}200)t] \end{aligned} \tag{2.11}$$

as depicted graphically in Figure 2.22.

If several targets with various doppler shifts were present, V_r would equal the sum of all these signals:

$$V_r = A_1 \sin [2\pi(f_o + f_{1d})t] + A_2 \sin [2\pi(f_o + f_{2d})t] + \ldots$$

where A_1, A_2, . . . , are the amplitudes and f_{1d}, f_{2d}, . . . , are the corresponding doppler shifts. Except for additional target returns, there will be no change in the described analysis.

The signal in Eq. (2.11) is then reduced through the local oscillator and the mixer of Figure 2.19 to an intermediate frequency f_{IF}. Assuming f_{IF} to be 30 MHz, we can write

$$\begin{aligned} V'_r &= \sin [2\pi(f_{\mathrm{IF}} + f_d)t] \\ &= \sin [2\pi(30{,}006{,}200)t] \end{aligned} \tag{2.12}$$

This signal is shown graphically in Figure 2.23. For digital signal processing, to accommodate a lower sampling rate, the signal is further

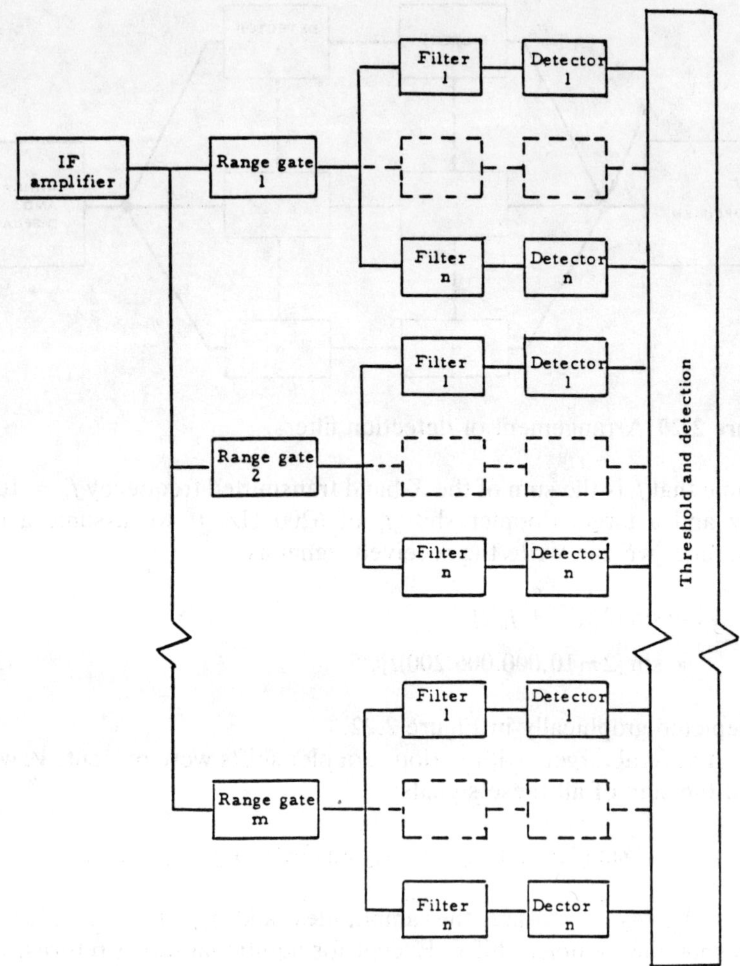

Figure 2.21 Range gates followed by doppler filters.

reduced in frequency. Assuming that a reduction to dc level is effected, we have

$$V_r'' = \sin\,[2\pi(f_{dc} + f_c)t] \tag{2.13}$$

Since $f_{dc} = 0$, we have

$$F = \sin\,[2\pi(6200)t] \tag{2.14}$$

as shown in Figure 2.24. The notation is changed from V to F in eq. (2.14) to conform to the standard Fourier transform notation. At this point, the signal contains the doppler shift, f_d, *only*. We are now ready to prepare the signal for the digital computer and to compute its frequency contents.

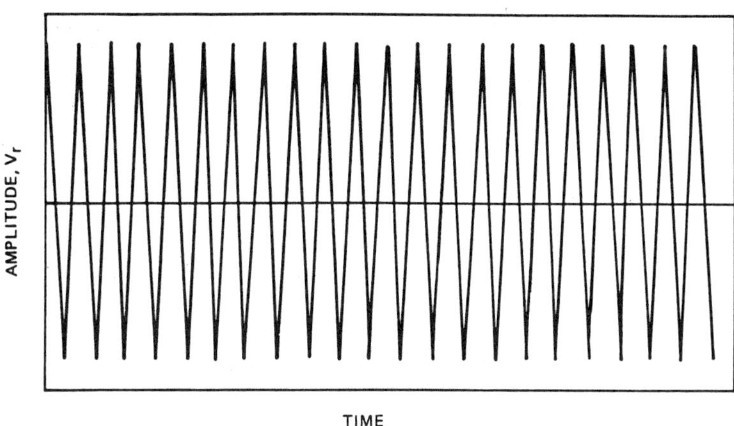

Figure 2.22 Received signal at frequency $f_o + f_d$.

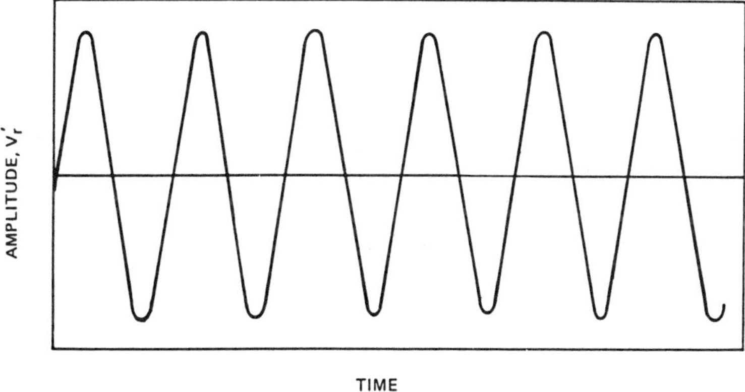

Figure 2.23 Received signal reduced to IF with frequency $f_{IF} + f_d$.

First, we select the total time span, T, used to sample the signal, as shown in Figure 2.24. This time span also determines a frequency

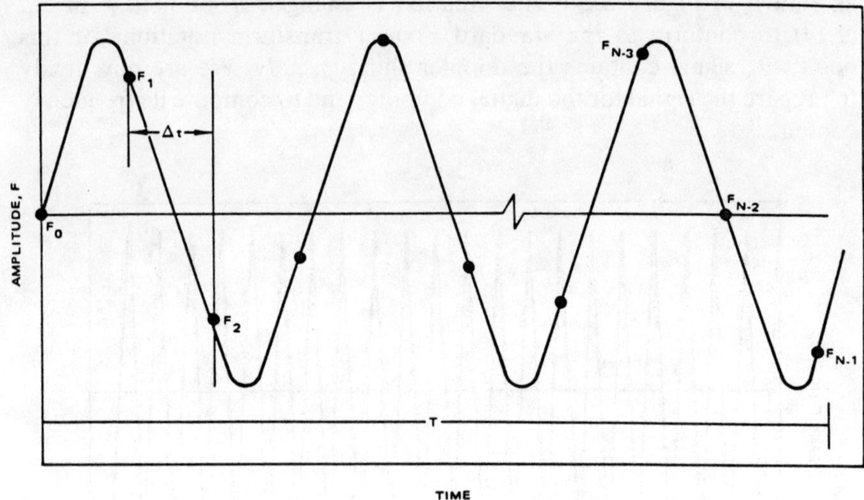

Figure 2.24 Received signal consisting of only the doppler shift, f_d.

step, Δf, which is equivalent to the filter bandwidths of the analog system of Figure 2.20. Referring to Figure 2.24, let us assume a time span, T, of 5 ms. From the relation

$$\Delta f = \frac{1}{T} \tag{2.15}$$

we compute a digital filter bandwidth of 200 Hz.

Assume that we can obtain adequate velocity coverage (doppler shift) with 250 filters. If we use binary arithmetic for computational efficiency (see [1], Chapter 9), the closest number to this will be $N = 2^8 = 256$. Our sampling interval Δt of Figure 2.24 then becomes

$$\Delta t = \frac{T}{N} = 19.53125\ \mu\text{s}$$

At this point, the analog signal of Figure 2.24 is sampled and converted by means of an analog to digital converter (A/D) to a set of numbers as given in Table 2.2. The time samples of this table are used in the Fourier transform equation

$$G_k = \sum_{n=0}^{N-1} F_n \exp\left[-(2\pi \mathrm{j}/N)(n)(k)\right] \tag{2.16}$$

Table 2.2 Time Samples of the Signal of Figure 2.24

Time	*Amplitude*
0	F_0
Δt	F_1
$2\Delta t$	F_2
⋮	⋮
$n\Delta t$	F_n
⋮	⋮
$(N-1)\Delta t$	F_{N-1}

where G_k is the amplitude of the signal having a frequency $k\Delta f$. Using this equation, the digital computer calculates the values of $G_o \ldots, G_k, \ldots, G_{N-1}$ as tabulated in Table 2.3. This computation is performed by replacing the exponential function of Eq. (2.16) by the trigonometric function:

$$\exp[-(2\pi \mathrm{j}/N)(n)(k)] = \cos(2\pi/N)(n)(k) - \mathrm{j}\sin(2\pi/N)(n)(k) \tag{2.17}$$

For example, with Eqs. (2.16) and (2.17), for a value of $k = 3$, the computer will use the following summation to calculate G_3:

$$\begin{aligned} G_3 &= F_o[\cos(2\pi/N)(0)(3) - \mathrm{j}\sin(2\pi/N)(0)(3)] \\ &\quad + F_1[\cos(2\pi/N)(1)(3) - \mathrm{j}\sin(2\pi/N)(1)(3)] \\ &\quad + \ldots + F_{N-1}[\cos(2\pi/N)(N-1)(3) \\ &\quad - \mathrm{j}\sin(2\pi/N)(N-1)(3)] \\ &= a_3 + \mathrm{j}\, b_3 \end{aligned} \tag{2.18}$$

where a_3 and b_3 are the real and the imaginary parts of G_3. The amplitude of G_3 then becomes

$$|G_3| = \sqrt{(a_3^2 + b_3^2)} \tag{2.19}$$

where the fences represent the absolute values. The value of $|G_3|$ represents the amplitude of the signal with a frequency $3\ \Delta f$. Equation (2.16) demonstrates the brute force method of calculating the values of G_k. Several computer subroutines are now available that carry out this computation more efficiently. These subroutines usually come under the general heading of *fast Fourier transform* (FFT) (see [4]).

Table 2.3 Output of Fourier Transform

Frequency	*Amplitude*
0	G_0
Δf	G_1
$2\Delta f$	G_2
$\vdots$	$\vdots$
$k\Delta f$	G_k
$\vdots$	$\vdots$
$(N-1)\Delta f$	G_{N-1}

In our example, we may use Eq. (2.16) and compute values of G_0, G_1, G_2, . . . , G_{255}. All of these G values will be zero except for G_{31}, corresponding to the frequency $f_d = 6200/200 = 31\ \Delta f$ as shown in Figure 2.25, since no other doppler shift, except $f_d = 6200$, is present in the original signal of Eq. (2.11).

The presence and the magnitude of a signal at a given frequency are determined by the value of its Fourier coefficient G; thus, by using digital signal processors, we can replace the analog filters and related equipment with a digital computer that evaluates the Fourier transform of the input signal.

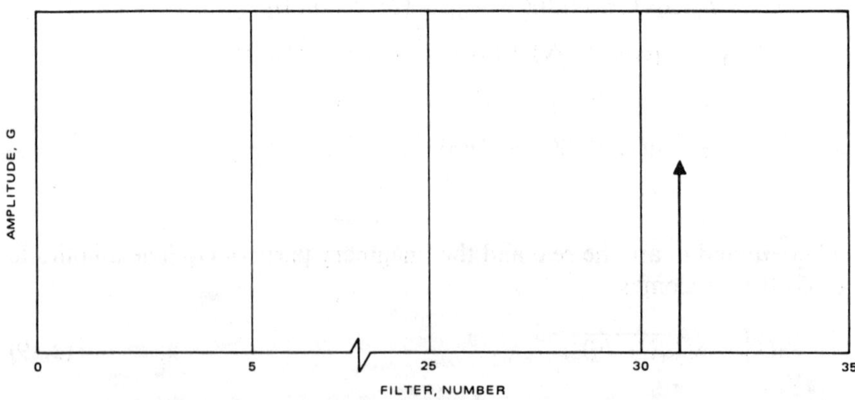

Figure 2.25 Presence of a signal in filter no. 31.

2.9 RADAR S/N RATIO EQUATION

The maximum detection and tracking range of a radar system is primarily a function of three parameters: transmitted power, antenna gain, and receiver sensitivity. Increasing the transmitted power will increase the radiated energy that, in turn, will result in a stronger target return. The antenna gain is a measure of the radiated energy in the direction of the target as compared to uniform radiation of energy. Receiver sensitivity is a measure of the capability of the receiver to detect target returns. When the power radiated by the radar is denoted as P, and this power is assumed to radiate uniformly in all directions, the power per unit area (or the power density) will be

$$\text{Power density} = \frac{P}{4\pi R^2} \tag{2.20}$$

where $4\pi R^2$ represents the area of a sphere with radius R. Since most radars have a directive antenna (that is, power is radiated in a specific direction), this is accounted for by incorporating the antenna gain in the equation. The antenna gain at a particular angle, Θ, is defined as the ratio of the radiated intensity at Θ to the radiation intensity of a uniformly radiating antenna (see Figure 2.26). At $\Theta = 0$, the maximum gain of an antenna is related to its physical area, A, by the equation:

$$G = \frac{4\pi A}{\lambda^2} \tag{2.21}$$

where A is the aperture area, and λ is the wavelength of the transmitted wave (see section 2.15). In Eq. (2.21), an antenna efficiency of 100 percent is assumed. When the antenna gain, G, is incorporated into Eq. (2.20), the power density in a particular antenna pointing direction is obtained:

$$\begin{matrix}\text{Radiated power density} \\ \text{of directive antenna}\end{matrix} = \frac{PG}{4\pi R^2} \tag{2.22}$$

The target intercepts and reflects part of the radiated energy. The portion of the power reflected by the target in the direction of the receiving antenna will be a function of the ability of the target to "focus" the

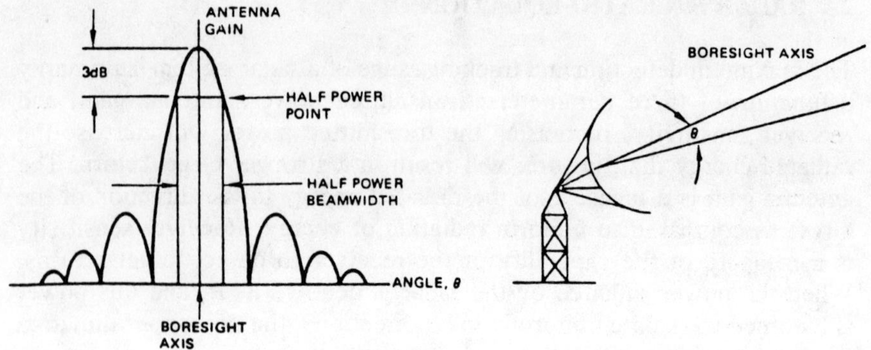

Figure 2.26 Typical antenna gain curves.

received energy in that direction. Qualitatively, this focusing ability is proportional to the target radar cross section, σ, which has the units of area (m^2). Using this, the reflected power from the target becomes

$$\text{Reflected power from the target} = \frac{PG\sigma}{4\pi R^2} \tag{2.23}$$

When the reflected power given by this equation is used as a uniform source of power located at the target, the power density of this energy at the radar (which is distance R away) can be calculated as follows:

$$\text{Power density of reflected target power at the radar} = \frac{PG\sigma}{(4\pi R^2)^2} \tag{2.24}$$

If we assume that the area capturing this energy is equal to the aperture area of the transmitting antenna, A (it is assumed that both transmitting and receiving antennas are one and the same):

$$\text{Power of the reflected signal at the radar} = \frac{PG\sigma A}{(4\pi R^2)^2} \tag{2.25}$$

Solving Eq. (2.21) for A and substituting in Eq. (2.25), we get

$$P_r = \frac{PG^2\lambda^2\sigma}{(4\pi)^3R^4} \tag{2.26}$$

where P_r designates target power received at the radar.

In the detection circuitry of the radar, this target return power will be processed in the presence of receiver noise. Receiver noise can originate within the receiver itself, or it may enter the receiver through the receiving chain, starting with the antenna and followed by the waveguides, amplifiers, and so on. One portion of the noise generated within the receiver is due to the thermal agitation of electrons in the receiver. This thermal noise, as given later, is proportional to the temperature, T, in Kelvins, and the receiver noise bandwidth, B_n, in Hz:

$$\text{Thermal noise} = kTB_n \tag{2.27}$$

where k is the Boltzmann's constant and equal to 1.38×10^{-23} J/deg. Equation (2.27) represents only one portion of the receiver noise; to account for all of the receiver noise, Eq. (2.27) is multiplied by an experimentally determined constant, called the noise figure and designated by F. With this factor, the total receiver noise can be written

$$N = FkTB_n \tag{2.28}$$

where N represents the total receiver noise.

Denoting the target power received by the radar [Eq. (2.26)] by S and the noise power [Eq. (2.28)] by N, we obtain the S/N ratio:

$$\text{S/N} = \frac{PG^2\lambda^2\sigma}{(4\pi)^3R^4(FkTB_n)} \tag{2.29}$$

Equation (2.29) represents the radar equation in its simplest form. Note from this equation that target signal power is inversely proportional to the fourth power of range and directly proportional to the square of antenna gain.

Equation (2.29) was derived without incorporating losses of energy that accompany transmission, reception, and the processing of electromagnetic radiation. Incorporating all of these losses in one term, we can write Eq. (2.29) as follows:

$$\text{S/N} = \frac{PG^2\lambda^2\sigma}{(4\pi)^3R^4(FkTB_n)L} \tag{2.30}$$

where L is the loss factor representing the sum total of transmit and receive losses.

2.9.1 Other Forms of the Radar Equation

If the transmit and receive antennas are not at the same location and do not have the same antenna gain, Eq. (2.30) can be written as

$$\mathrm{S/N} = \frac{PG_tG_r\lambda^2\sigma}{(4\pi)^3R_t^2R_r^2(FkTB_n)L} \tag{2.31}$$

where R_t and R_r are ranges between the target and the transmit antenna and the target and the receive antenna, respectively. The corresponding transmit and receive antenna gains are G_t and G_r.

In cases where a signal is transmitted from one radar and is received by another radar, the beacon (rather than the radar) equation applies. The beacon equation can be readily derived from the previously given information as follows:

$$\mathrm{S/N} = \frac{PG_tG_r\lambda^2}{(4\pi)^2R^2(FkTB)_rL} \tag{2.32}$$

where

R = range between transmitting and receiving radars
G_t = transmit antenna gain
G_r = receive antenna gain
$(FkTB)_r$ = receiver noise level

Note that in the case of Eq. (2.32), the S/N ratio is proportional to the second (rather than the fourth) power of range R.

2.9.2 Integration of Returned Pulses

Note that Eq. (2.30) was derived for a single returned pulse. In search radars where the antenna sweeps across the target, depending on the rate of sweep and the number of transmitted pulses, a great number of pulses can be integrated to improve the attained S/N ratio. Assuming the number of transmitted pulses per second to be designated by PRF and the on-target time by T_i, the number of target returned pulses can be written:

$$N_p = T_i \cdot \mathrm{PRF} \tag{2.33}$$

where N_p is the number of returned pulses. Using Eq. (2.33) in S/N ratio Eq. (2.30), we get

$$\text{S/N} = \frac{PG^2\lambda^2\sigma T_i\,\text{PRF}}{(4\pi)^3R^4(FkTB_n)L} \tag{2.34}$$

For a transmitted pulsewidth of τ, the matched receiver filter bandwidth, B_n, is given by $1/\tau$ (see Figure 2.27). Using this in Eq. (2.34), we get

$$\text{S/N} = \frac{PG^2\lambda^2\sigma T_i\,\text{PRF}\,\tau}{(4\pi)^3R^4(FkT)L} \tag{2.35}$$

From Figure 2.27, note that $P\tau$ represents energy transmitted per pulse. Dividing this energy by the duration of the interpulse period, $\overline{T}$, we can compute the average power:

$$\begin{aligned} P_{\text{AVE}} &= \frac{P\tau}{\overline{T}} \\ &= P\tau\,\text{PRF} \end{aligned} \tag{2.36}$$

where P_{AVE} is the average power, and $\overline{T} = 1/\text{PRF}$ is the interpulse period. Using Eq. (2.36) in Eq. (2.35), we get

$$\text{S/N} = \frac{P_{\text{AVE}}T_iG^2\lambda^2\sigma}{(4\pi)^3R^4(FkT)L} \tag{2.37}$$

Equation (2.37) represents the S/N ratio of a matched transmit-receive radar system. From Eq. (2.37), note that the S/N ratio, as expected, is proportional to the product of the average power by the on-target time, which represents the amount of transmitted energy. As shown in [1], Eq. (2.37) applies to all types of radar: low, medium, and high PRF systems.

2.10 SEARCH RADARS

Two parameters, scan coverage and detection range, dominate the design of search radars. In this section, starting with the S/N ratio Eq. (2.37), parametric equations are derived describing the interrelation of radar parameters with these factors.

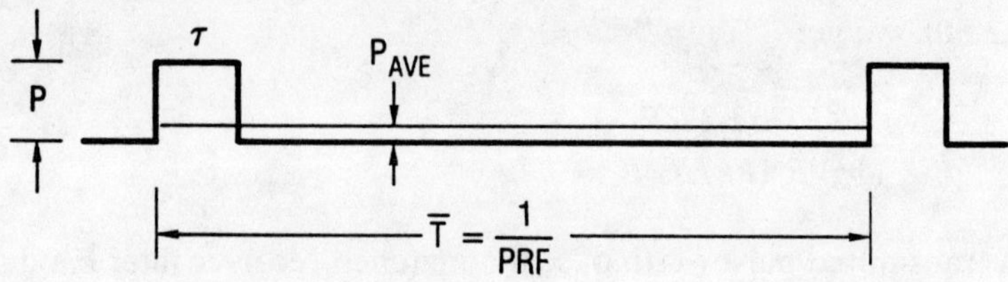

Figure 2.27 Transmitted pulse τ, interpulse period T, and average power.

Using the antenna gain equation in the form

$$G = \frac{4\pi A}{\lambda^2} \tag{2.38}$$

where A is the aperture area and λ is the wavelength, in Eq. (2.37), we get

$$\text{S/N} = \frac{P_{\text{AVE}} A^2 \sigma T_i}{(4\pi)\lambda^2 R^4 (kT) L} \tag{2.39}$$

Assuming a circular aperture with diameter D and substituting the value of $\pi D^2/4$ for one of the areas A in Eq. (2.39), we get

$$\text{S/N} = \frac{P_{\text{AVE}} D^2 A \sigma T_i}{(16)\lambda^2 R^4 (kT) L} \tag{2.40}$$

For an angular scan coverage of Ω sr and an antenna linear beamwidth of Θ_{BW}, we can calculate the number of antenna beam positions, n_p, within the scan coverage:

$$n_p = \frac{\Omega}{\Theta_{\text{BW}}^2} \tag{2.41}$$

where Θ_{BW}^2 represents the antenna beamwidth in sr. Using the relation $\Theta_{\text{BW}} = \lambda/D$, where the beamwidth is represented as a function of wavelength λ and diameter D, in Eq. (2.41) we get

$$n_p = \frac{D^2}{\lambda^2}\Omega \tag{2.42}$$

For a scan duration or radar frametime of t_s s, using Eq. (2.42) we can obtain the on-target time, T_i, as

$$T_i = \frac{t_s}{n_p} = \frac{\lambda^2}{D^2}\frac{t_s}{\Omega} \tag{2.43}$$

Substituting Eq. (2.43) into Eq. (2.40) for T_i, we get

$$\mathrm{S/N} = \frac{P_{\mathrm{AVE}} A \sigma}{(16) R^4 (kT) L} \cdot \frac{t_s}{\Omega} \tag{2.44}$$

Equation (2.44) gives the S/N ratio for a search radar with scan coverage of Ω sr and radar frame time (or scan time) of t_s s. Equation (2.44) is independent of radar transmit frequency. For a given set of radar parameters and scan coverage, the S/N depends on power-aperture product. Thus, for a given power-aperture product, Eq. (2.44) can be solved for range, R, in terms of radar parameters:

$$R = \left[\frac{P_{\mathrm{AVE}} A \sigma}{(16)(kT) L (\mathrm{S/N})} \cdot \frac{t_s}{\Omega}\right]^{1/4} \tag{2.45}$$

From Eq. (2.45) note that the detection range of a search radar is inversely proportional to the area coverage, Ω, and directly proportional to radar scan time, t_s. The value of S/N of Eq. (2.45) represents that required for target detection. Observe further that the detection range of the search radar as given by Eq. (2.45) is independent of transmitting radar frequency.

2.11 TRACKING RADARS

Tracking radars "memorize" the past spatial positions of the target and predict future positions based on these past data; that is, they keep target tracks. Target tracking can be either continuous or discontinuous. In continuous tracking the radar beam is kept on the target and information is extracted to predict future target positions continuously. Discrete tracking uses periodic and noncontinuous radar data (such as obtained in search radars) and tracks the targets in a computer. These radars, usually called track-while-scan radars, scan a given search volume and illuminate several targets, periodically extracting space angles (azimuth and elevation), range, and range rate data from each target. This information is then processed in a digital computer to accurately calculate the spatial positions of the targets. For a complete discussion of multiple target radar data processing, see [5]. Several signal processing methods are used in continuous tracking radars: sequential lobing, conical scan tracking, and amplitude-comparison monopulse are a few examples.

2.11.1 Sequential Lobing and Conical Scan Tracking

In this method, the antenna beam is switched between two positions in each vertical and horizontal plane. In each position, target strength is measured and converted to a voltage. The difference between these voltages is a measure of angular error between the actual position of the target and the position of the antenna lobing axis when the measurement was made. Figure 2.28a shows an example of this method in a single plane.

Note from this figure that the target is illuminated at a higher point of antenna gain pattern in Position 1 than in Position 2. Moving the lobing axis to Position 1 and repeating the process will eventually reduce the error and put the lobing axis on the target direction. More sophisticated servo systems can be designed to utilize the magnitude and direction of the difference voltage between Positions 1 and 2 of the lobe and converge the lobing axis to actual target position more rapidly.

Conical scan tracking is the logical extension of sequential lobing where the antenna beam is rotated about the axis, as shown in Figure 2.28b. In this method, the target return is examined during one complete rotation of the antenna boresight axis around the axis of rotation. Using these data, azimuth and elevation commands are generated by a servo system to make the axis of rotation of the antenna colinear with target direction.

2.11.2 Monopulsed Tracking

In the continuous tracking methods just described, more than one target return pulse is required to establish radar tracking error. In practice, the number of consecutive pulses is four, since two pulses are required in each of the vertical and horizontal directions. In these methods, fluctuation of target size between pulses results in tracking errors. The target size–fluctuation problem can be eliminated if a single pulse is used to determine

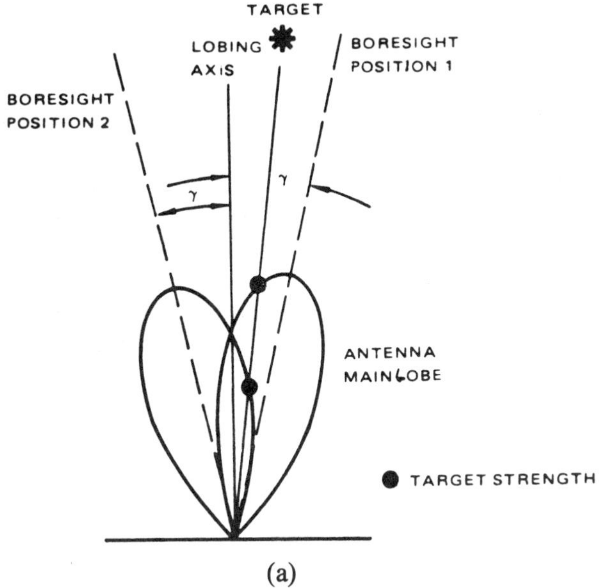

(a)

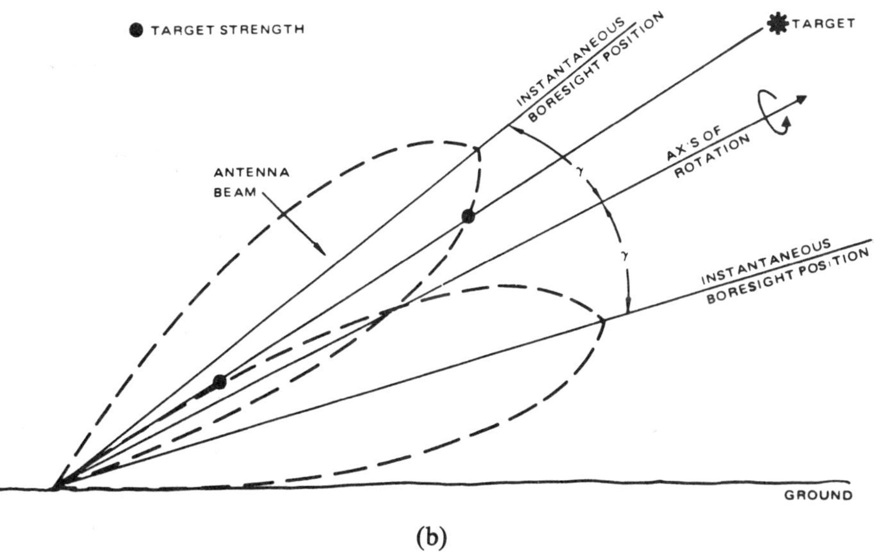

(b)

Figure 2.28 Single target tracking antennas: (a) sequential lobing; (b) conical scan.

tracking error. Tracking systems that use a single pulse to calculate tracking errors are called, appropriately, monopulse systems (see [6]). Angle errors in these systems are derived by using either amplitude or phase comparison of returned signals.

Figure 2.29 illustrates the basics of the amplitude-comparison monopulse system. Two overlapping antenna beams are needed for each of the two coordinate directions (azimuth and elevation). These beams may be generated with a single reflector illuminated by four adjacent feeds. The sum pattern is used for transmission, while both the sum and the difference patterns are used for reception. The sum channel is used for range measurement and as a reference to obtain the sign of the angular error. Figure 2.30 gives typical sum and difference gain curves as a function of the angle off boresight, Θ. For small angles ($\Theta < 2°$), the ratio of relative amplitudes of target returns from sum and difference channels will result in a voltage proportional to angular error, as illustrated in Figure 2.30.

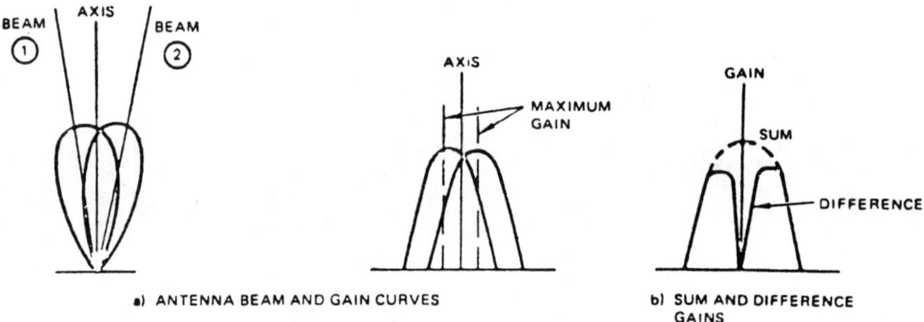

Figure 2.29 Amplitude-comparison monopulse sum and difference beams (a) and gain (b) patterns.

Figure 2.31 shows a block diagram of a low PRF amplitude-comparison monopulse radar for a single (azimuth or elevation) channel. The duplexer switches the antenna between transmit and receive functions. Note that the transmission is through the sum channel. The mixers of the figure reduce the target received frequency from *radio frequency* (RF) to *intermediate frequency* (IF). The phase-sensitive detector of the figure determines the sign of the angular error, right and left for azimuth channel and up and down for elevation channel. The final result is the detection of the range and angular error signal as shown.

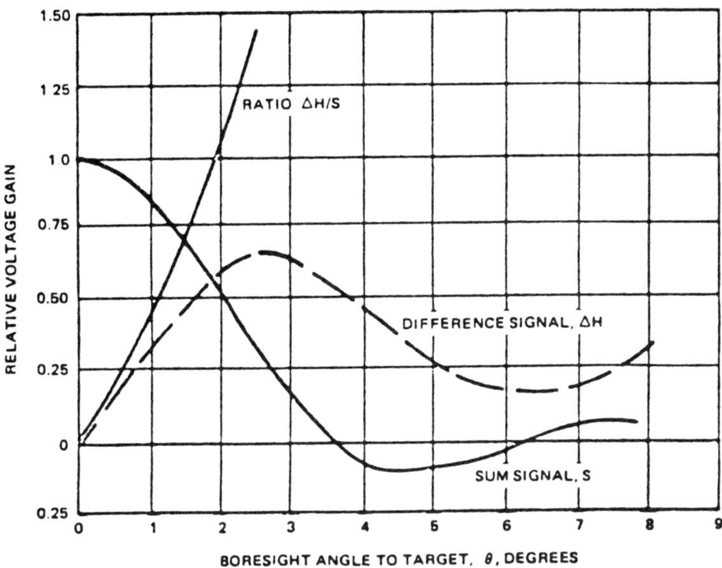

Figure 2.30 Sum and difference gain patterns and angular tracking error in amplitude-comparison monopulse.

2.11.3 Track-while-Scan Radars

Modern day scanning radars are capable of obtaining target observations (range and angular position) for several targets within their scan illumination coverage. These radars can be used for multiple target tracking and airspace surveillance [5].

Figure 2.32 shows a typical inertial coordinate system with the radar at the origin. The north and the east axes point to the north and east, respectively; the vertical axis is perpendicular to these axes. Denoting the respective direction cosines by N, E, and V, we define the position of each target with respect to the origin by range, R, and two out of three direction cosines. Target angular position can also be specified in terms of azimuth, η, and elevation, ϵ, angles rather than the N, E, and V as just described.

For a scanning radar, the target data rate will depend on the radar frame time (time between consecutive illuminations of a specific point in space). Thus, tracking of several targets becomes largely a matter of

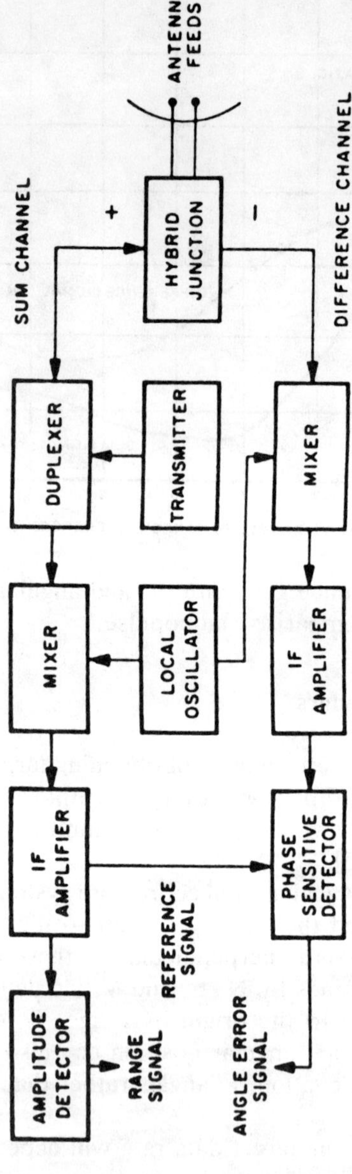

Figure 2.31 Amplitude-comparison monopulse radar block diagram.

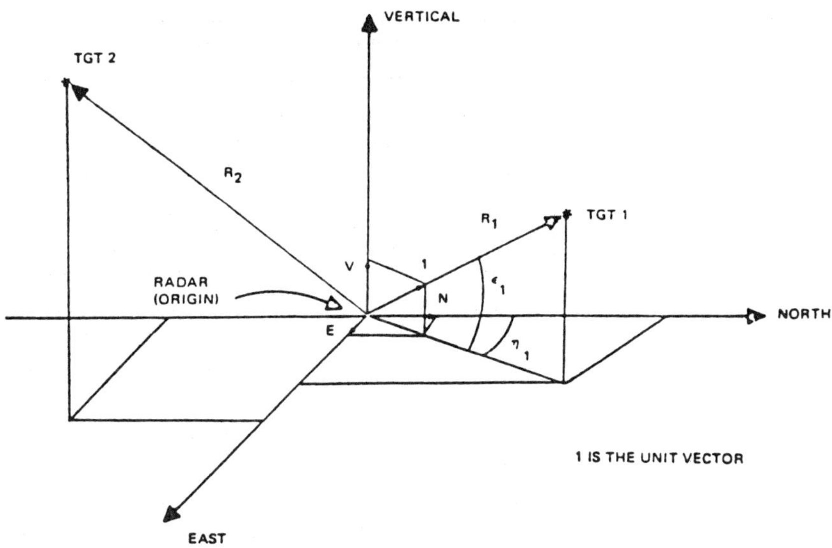

Figure 2.32 North, east, and vertical coordinate system.

sorting target observations obtained in each radar frame and assigning them to proper target tracks. These tracks form the basis for estimating future target positions. The process of tracking targets based on discrete radar information obtained while the radar continues to scan the airspace is referred to as *track-while-scan* (TWS) and takes place in a digital computer.

Figure 2.33 illustrates the data processing required for the computer's TWS operation. The angular position of each radar target observation is converted to the direction cosines (N, E, V) of the inertial coordinate system.

This inertial angular position, together with range R, specifies the inertial target position, RNEV, Block 2 in the figure. Note that although one of the direction cosines is redundant, all three are carried as target coordinates to ensure maximum tracking accuracy. These target observations are correlated and associated with existing tracks. The correlation and association process ensures that proper observations are assigned to each target track by comparing these observations with the predicted target positions. Subsequently (Block 4), these target observations are

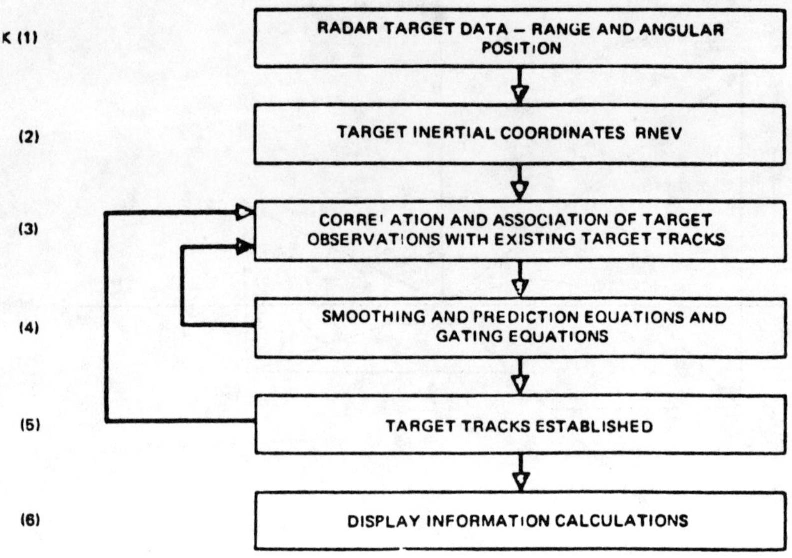

Figure 2.33 TWS data processing.

incorporated in the proper target tracks through smoothing and prediction equations that primarily use Kalman filtering methods (see Figure 2.34). For each radar frame, the acceptability region for correlating and associating target observations (the gates) is established around the predicted target positions for each of the four coordinates, R and NEV. At Block 5, target tracks, together with predicted target positions for the next radar frame and the appropriate gates, are calculated and stored within the computer. This information is used by Blocks 3 and 4 for correlation, association, and gating. It may also be used as the basis for computing information to be displayed (Block 6).

2.12 MEASUREMENT AND TRACKING ACCURACY

An important consideration in search and tracking radars is the accuracy of the radar measurements and the accuracy with which targets are tracked. Measurement accuracy is primarily a function of radar characteristics such as range gate duration, doppler filter bandwidth, antenna beamwidth, and radar signal processing methods. Tracking accuracy, on the other hand, is a function of the amount of data processing and smoothing performed on the measured data. Present-day radar systems use a variety of filtering and smoothing methods on measured data to

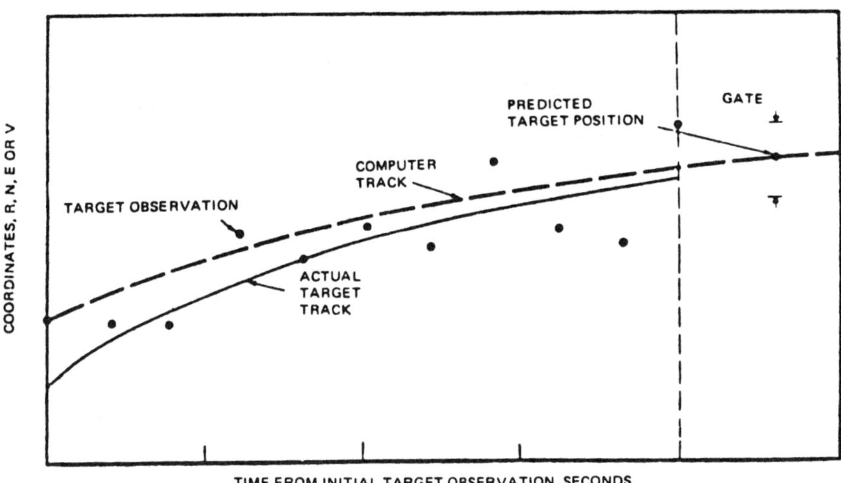

Figure 2.34 Smoothing and prediction of target data.

obtain desired target tracking accuracy. These smoothing methods are typified by least squares fit techniques and Kalman filtering methods. In this section, however, we shall concentrate only on radar measurement accuracies.

In low PRF radars, range measurement accuracy is primarily determined by the size of the range gate. In most cases, a typical target return will fall into two adjacent range gates, and simple averaging of the corresponding range values of each gate will result in a range accuracy of

$$\Delta R = \frac{c\tau}{4} \tag{2.46}$$

where ΔR is the range measurement accuracy, c is the velocity of light, and τ is the pulse length (range gate duration).

Similarly, the accuracy with which a doppler radar can measure radar-target range rate, $\dot{R}$, will depend on the frequency separation of two adjacent filters (doppler filter bandwidth). When this separation is denoted by Δf_s, the range rate measurement accuracy becomes

$$\Delta \dot{R} = \frac{\lambda \Delta f_s}{4} \tag{2.47}$$

where $\Delta\dot{R}$ is the range rate measurement accuracy, λ is the wavelength, and Δf_s is the doppler filter bandwidth.

The accuracy of radar angular measurements is a function of the antenna half-power beamwidth, Θ_{BW}. For simply constructed antennas, the antenna beamwidth can be approximated as a function of wavelength and antenna size as follows:

$$\Theta_{BW} = \lambda/l\,(\text{rad})$$
$$\simeq \frac{60\lambda}{l}\,(\text{deg}) \tag{2.48}$$

where Θ_{BW} is the half-power antenna beamwidth, λ is the wavelength, and l is the aperture dimension. In a circular antenna, l will equal the diameter.

Continuous tracking radars, as shown in Figures 2.28a and 2.28b, can always keep the target within one-half of their antenna beamwidths. Thus, the angular measurement accuracy can be written as

$$\Delta\Theta = \frac{\Theta_{BW}}{2} \tag{2.49}$$

where $\Delta\Theta$ is the angular measurement accuracy and Θ_{BW} is the antenna beamwidth. In search radars, on the other hand, the angular measurement accuracy is usually twice the value given in Eq. (2.49).

Application of sophisticated signal processing methods and waveform analysis will show that (1) the behavior of measurement accuracies given by Eqs. (2.46)–(2.49) most accurately follow a Gaussian distribution, and (2) the limiting value of measurement accuracy is a function of the S/N ratio. Using these two effects in Eqs. (2.46)–(2.49), we can write typical measurement accuracies as

$$\sigma_R = \frac{c\tau}{4\sqrt{\text{S/N}}} \tag{2.50a}$$

$$\sigma_{\dot{R}} = \frac{\lambda \Delta f_s}{4\sqrt{\text{S/N}}} \tag{2.50b}$$

and

$$\sigma_\Theta = \frac{\Theta_{BW}}{2\sqrt{\text{S/N}}} \tag{2.50c}$$

where σ_R, and $\sigma_{\dot{R}}$, and σ_Θ are the standard deviations of range, range rate, and angular measurements, respectively. The square root of the S/N power ratio, representing the S/N voltage ratio, is used in Eq. (2.50). Note that these accuracies are obtained from a consideration of signal processing and thermal noise characteristics and do not include errors from other measurement sources, such as servo systems, quantization, *et cetera*.

2.13 DESIGN EXAMPLE

2.13.1 Ground-Based Low PRF Radar

Assume that there is a requirement to design a ground-based, low PRF radar with the capability of tracking a 5 m^2 target at a range of 40 nmi. The design may begin by a study of existing radars of similar capability [7] and development of parameters for the required radar system. Using this procedure, we may arrive (after several iterations) at the following parameters:

Transmit frequency, f	=	6000 MHz (C-band)
Pulse repetition frequency, PRF	=	1500 Hz
Antenna diameter, l	=	4 ft
Receiver noise figure, F	=	4
Noise temperature, T	=	300 K
Transmit power, P	=	100 kW
Pulse length, τ	=	0.5 μs
Receiver noise bandwidth, B	=	2 MHz
System losses, L	=	8 dB
	=	6.3095
Target tracking, S/N	=	4 dB
	=	2.5119

In this, a matched receiver bandwidth is selected. This bandwidth is approximately equal to $1/\tau$, where τ is the transmitted pulse length. From previous equations, we can calculate the antenna gain and wavelength as follows:

$$G = \frac{4\pi A}{\lambda^2}; \quad \lambda = c/f$$

$$G = 5.868 \times 10^3; \quad \lambda = 0.164 \text{ ft}$$

$$= 37.685 \text{ dB} \tag{2.51}$$

Using Eq. (2.30), we can calculate the corresponding S/N:

$$\begin{aligned} \text{S/N} &= \frac{(100 \times 10^3)(5.868 \times 10^3)^2(0.164)^2(5)(3.2808)^2}{(4\pi)^3(R^4)(4)(1.38 \times 10^{-23})(300)(2 \times 10^6)(6.3095)} \\ &= \frac{1.202 \times 10^{22}}{R^4} \end{aligned} \tag{2.52}$$

where 3.2808 is the conversion factor from meters to feet. Since a S/N ratio of 4 dB or 2.5119 is required for target tracking, the target tracking range will be

$$R = \left(\frac{1.202 \times 10^{22}}{2.5119}\right)^{1/4} = 2.6301 \times 10^5 \text{ ft} = 43.25 \text{ nmi} \tag{2.53}$$

where a conversion factor of 6080 ft/nmi is used. Note, therefore, that this radar-target tracking range is close to the requirement of 40 nmi.

The PRF of 1500 Hz will result in an unambiguous range of

$$\begin{aligned} R_{\text{ua}} &= \frac{c}{2\,\text{PRF}} \\ &= \frac{3 \times 10^8 \times 3.2808}{2 \times 1500 \times 6080} \\ &= 53.96 \text{ nmi} \end{aligned} \tag{2.54}$$

which is in excess of 40 nmi tracking range and is acceptable.

Consider this use of the radar parameters in the search configuration and assume that we want to implement a 360 × 10° azimuth and elevation search in a 10-s radar frame time. Using the search radar Eq. (2.45)

$$R = \left[\frac{P_{\text{AVE}} A\sigma}{(16)(FkT)(\text{S/N})L} \cdot \frac{t_s}{\Omega}\right]^{1/4} \tag{2.55}$$

we can compute the detection range of this search radar. The terms of Eq. (2.55) can be computed as follows:

$$\begin{aligned} P_{\text{AVE}} &= P_{\text{T}}\,\text{PRF} \\ &= 75\text{W} \end{aligned}$$

$$A = \frac{\pi D^2}{4}$$
$$= 1.168 \text{ m}^2$$
$$\Omega = \frac{360 \times 10}{(57.3)^2}$$
$$= 1.096 \text{ sr} \tag{2.56}$$

Additionally, assume that a S/N of 15 dB (a factor of 31.62) is required for target detection; from Eq. (2.55) we have

$$R = \left[\frac{(75)(1.168)(5)}{(16)(1.38\text{E} - 23)(4)(300)(31.62)(6.3095)} \cdot \frac{10}{1.096}\right]^{1/4}$$
$$= 93.24 \text{ km}$$
$$= 50.30 \text{ nmi} \tag{2.57}$$

Observe that the detection range of this radar is slightly less than its unambiguous range corresponding to a PRF of 1500 pulse/s as given by Eq. (2.54), as it should be in a well-designed radar.

In this example, an S/N of 15 dB was assumed for detection. This value can be computed from more sophisticated analyses that include probability of detection, threshold setting, and target cross section fluctuations, as discussed at the end of this chapter.

Additional radar characteristics can also be obtained from the given date. Using the antenna diameter of 4 ft, we get the antenna beamwidth from Eq. (2.48):

$$\Theta_{BW} = \frac{(60)(0.164)}{4} = 2.46° \tag{2.58}$$

From Eq. (2.49), the angular measurement accuracy will be 1.23°—one-half the antenna beamwidth value as given in Eq. (2.58). The range measurement accuracy of a low PRF radar is given by Eq. (2.46) as

$$\Delta R = \frac{3 \times 10^8 \times 3.2808 \times 0.5 \times 10^{-6}}{4}$$
$$= 123 \text{ ft} \tag{2.59}$$

2.13.2 Pulsed Doppler Radar

As a variation of this design, consider a high PRF pulsed doppler radar with applicable parameters of the radar in subsection 2.13.1 that uses frequency modulation ranging and a doppler filter beamwidth of 1000 Hz. The list of parameters of this radar follows (an asterisk indicates items changed from the previous low PRF radar design):

Transmit frequency, f	= 6000 MHz (C-band)
*Pulse repetition frequency, PRF	= 200,000 Hz
Antenna diameter, l	= 4 ft
Receiver noise figure, F	= 4
Noise temperature, T	= 300 K
*Transmit power, P	= 10 kW
*Pulse length, τ	= 2 μs
*Receiver noise bandwidth (assumed equal to doppler filter bandwidth), B	= 1000 Hz
System losses, L	= 8 dB
	= 6.3095
Target tracking, S/N	= 4 dB
	= 2.5119

The pulse repetition frequency PRF is increased from 1500 to 200,000 pulses/s, as the interpulse period is no longer used to obtain radar-target range. Because of increased PRF, the transmitted power is decreased from 100 to 10 kW to avoid transmitter burnout or power dumps. The pulse length is increased to two μs from its low PRF value of 0.5 μs, since, in the case of high PRF radar, the pulse length no longer determines radar-target range measurement accuracy. Note that a PRF of 200,000 corresponds to an interpulse period of 5 μs: 2 μs for transmission and 3 μs for reception. In the case of doppler radars, the effective transmitted peak power, called the centerline power, will be proportional to the square of the transmitter duty cycle: (2/5)Λ2. The receiver noise power will be proportional to the receiver duty cycle or 3/5.

The receiver noise bandwidth no longer needs to be matched to the transmitted pulse and can be decreased to 1000 Hz or less. The criterion is the availability of doppler filters and the cost of implementation. The doppler filter time constant, however, should be matched to the duration of the target energy return. Considering the effect of new parameters in the radar equation, we shall increase the target tracking range by the following factor:

$$\left[\frac{2 \times 10^6\,\text{Hz}}{1000\,\text{Hz}} \times \frac{10\,\text{kW}\,(2/5)\Lambda 2}{100\,\text{kW}\,(3/5)}\right]^{1/4} = 2.70 \tag{2.60}$$

where the last term accounts for transmit and receive duty cycles. Using this and Eq. (2.53), we note that the high PRF doppler radar will have a target tracking range of

$$43.25 \times 2.70 = 116.77\,\text{nmi} \tag{2.61}$$

As noted in the following discussion, this approximately threefold increase in tracking range is achieved at the expense of degraded range accuracy and sophisticated hardware design. Using frequency modulation ranging, we can obtain the range quantization ΔR from Eq. (2.4) as follows:

$$\Delta R = \frac{cB}{2\dot{f}} \tag{2.62}$$

where B is the doppler filter bandwidth, $\dot{f}$ is the modulation slope, and c is the velocity of light. For an assumed modulation slope of $\dot{f} = 10 \times 10^6$ Hz/s, Eq. (2.62) results

$$\begin{aligned}\Delta R &= \frac{3 \times 10^8 \times 3.2808 \times 1000}{2 \times 10 \times 10^6} \\ &= 4.9212 \times 10^4\,\text{ft} \\ &= 8.09\,\text{nmi}\end{aligned} \tag{2.63}$$

Thus, a high PRF doppler radar of equivalent design achieves higher target tracking ranges but at the expense of more sophisticated hardware design (doppler filters and transmitted frequency modulations) and degraded range accuracy. Note that this high PRF radar will have a range rate measurement accuracy, from Eq. (2.47), of

$$\begin{aligned}\Delta \dot{R} &= \frac{\lambda B}{4} \\ &= \frac{0.164 \times 1000}{4} \\ &= 41\,\text{ft/s}\end{aligned} \tag{2.64}$$

In a number of cases, this range rate accuracy is used to track targets in velocity.

2.14 ANTENNA GAIN PATTERN AND ELECTRONIC SCAN

2.14.1 Antenna Gain Pattern

The antenna gain pattern can be derived from basic relations by considering the propagation of electric field from a set of radiating elements. Consider the radiating elements of Figure 2.35, with each element radiating energy with the same amplitude and phase. At a point P in the plane of the figure, the total sum of the energy can be computed by obtaining the phase of emitted radiation from each element. Assuming the phase shift of the radiated energy from point zero (0) arriving at P to be zero, we have

$$
\begin{aligned}
\text{Phase shift of point 0 at } P &= 0 \\
\text{Phase shift of point 1 at } P &= R - R_1 \\
&= \mathrm{d} \sin \eta \\
&= \mathrm{d} \sin \eta \; 2\pi/\lambda \ (\mathrm{rad}) \\
\text{Phase shift of point 2 at } P &= R - R_2 \\
&= 2\mathrm{d} \sin \eta \\
&= 2\mathrm{d} \sin \eta \; 2\pi/\lambda \ (\mathrm{rad}) \\
\text{Phase shift of point n at } P &= R - R_n \\
&= n\mathrm{d} \sin \eta \\
&= n\mathrm{d} \sin \eta \; 2\pi/\lambda \ (\mathrm{rad})
\end{aligned}
$$

In these relations, one wavelength λ is set equal to 2π rad.

In general, field strength at P for N radiating elements, with each element radiating with an amplitude of unity, will be the sum of radiation from each element as follows:

$$
\mathrm{E}(\eta) = \sum_{n=0}^{N} \cos\left(\frac{2\pi}{\lambda} n\mathrm{d} \sin \eta\right) \tag{2.65}
$$

where N is the number of radiating elements and $\mathrm{E}(\eta)$ is the field strength at P. In Eq. (2.65), a cosine wave propagation is assumed.

For symmetrically located radiating elements as shown in Figure 2.36, the field strength $\mathrm{E}(\eta)$ becomes

$$\mathrm{E}(\eta) = \sum_{-N}^{N} \cos\left(\frac{2\pi}{\lambda} nd \sin\eta\right) \tag{2.66}$$

where the limits are from $-N$ to N. Normalizing the value $\mathrm{E}(\eta) = 1$ for $\eta = 0$, we get

$$\mathrm{E}(\eta) = \frac{1}{2N + 1} \sum_{-N}^{N} \cos\left(\frac{2\pi}{\lambda} nd \sin\eta\right) \tag{2.67}$$

where $1/(2N + 1)$ is the normalizing factor that will result in $\mathrm{E}(0) = 1$.

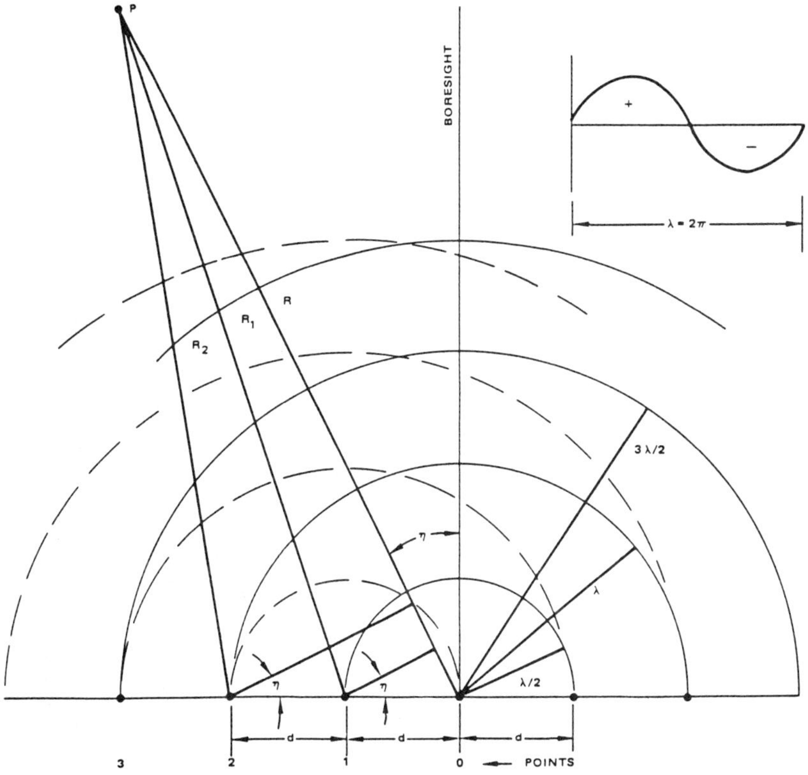

Figure 2.35 Equidistance radiating elements.

The summation of Eq. (2.67) for an antenna consisting of a continuous radiator can be written in the integral form. Inherent in going from

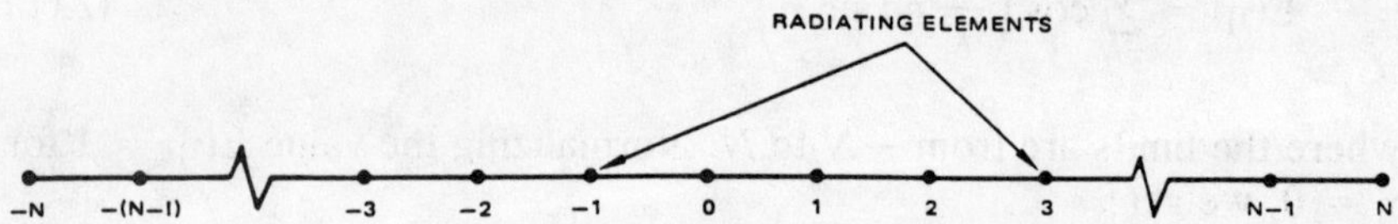

Figure 2.36 Arrangement of $2N + 1$ radiating elements.

discrete summation to continuous integral is the assumption that the radiating elements are closely spaced. Thus, the subject integral becomes

$$E(\eta) = \frac{1}{2N + 1} \int_{-N}^{N} \cos nx \, dn \tag{2.68}$$

where

$$x = \frac{2\pi}{\lambda} d \sin \eta \tag{2.69}$$

and the variable of integration is n. Integration of Eq. (2.68) results in

$$E(\eta) = \frac{1}{2N + 1} \frac{2 \sin Nx}{x} \tag{2.70}$$

For large N; $2N + 1 \simeq 2N$, and we get

$$E(\eta) = \frac{\sin Nx}{Nx} = \frac{\sin \alpha}{\alpha} \tag{2.71}$$

where Nx is equal to α. Equation (2.71) represents the often quoted sin x/x behavior of the antenna pattern for a linear array of radiating elements. Note that Eq. (2.71) represents voltage distribution; converting this to power, we get

$$G = [E(\eta)]^2 = (\sin \alpha/\alpha)^2 \tag{2.72}$$

where G is the antenna power gain. A plot of Eq. (2.72) is given in Figure 2.37. Note that the value of α in Eqs. (2.71) and (2.72) is directly related to the physical dimensions of the antenna. From previously given relations, we have

$$\alpha = Nx$$
$$= \frac{2\pi}{\lambda}(N\mathrm{d}) \sin \eta \qquad (2.73)$$

For an antenna of length l, Nd will be $l/2$ and Eq. (2.73) becomes

$$\alpha = \frac{\pi l}{\lambda} \sin \eta \qquad (2.74)$$

Thus, the parameter α in the antenna gain Eq. (2.72) is related to antenna length, angle off boresight, η, and transmitted wavelength, λ.

Figure 2.37 shows the half-power or 3-dB antenna beamwidth together with its sidelobe structure. Half-power beamwidth can be calculated as

$$\Theta = \frac{51\lambda}{l} \text{(deg)} \qquad (2.75)$$

where Θ is the half-power beamwidth and l is the antenna length. The first sidelobe in the case of Figure 2.37 is 13.2 dB down from the mainlobe.

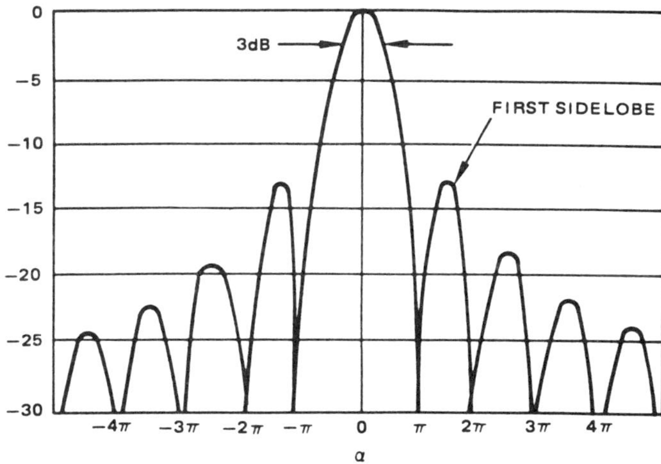

Figure 2.37 Antenna gain pattern $(\sin \alpha/\alpha)^2$.

This derivation of antenna gain pattern can be extended to two-dimensional cases, where radiating elements are arranged in a planar array. A special and interesting case of planar array antenna is the circular antenna. In this case, the radiation field intensity involves first order Bessel functions with the resulting half-power beamwidth

$$\Theta = \frac{58.5\lambda}{D} \text{ (deg)} \tag{2.76}$$

where D is the antenna diameter. In this case, the first sidelobe will be 17.6 dB (power) from the antenna mainlobe instead of 13.2 dB of sin α/ α radiation pattern.

2.14.2 Electronic Scan Antennas

The antenna gain pattern just discussed can be generated by a parabolic dish antenna with a feed horn or, in the case of planar array antennas, by a series of slots on the surface of the antenna that are backed by interconnected waveguides. These waveguides are constructed to produce uniform phasing of the radiated wavefront on the surface of the planar array antenna. In both cases, a parabolic dish with a feed horn and the planar array antenna, the antenna is moved mechanically through azimuth and elevation gimbals to spatially position its gain patterns. This movement, in effect, changes the position of a stationary gain pattern (with respect to the antenna dish) in order to obtain greater coverage. This coverage can also be obtained by keeping the antenna stationary and changing the position of the generated antenna gain pattern as is the case in electronic scanning antennas.

Electronic scanning antennas are mechanized by controlling the phase of generated waveform at each radiating element. By proper control of this phasing, the angular position of the antenna beam can be varied electronically as opposed to the mechanical motion of the antenna dish [8].

Figure 2.38 shows an antenna with five radiating elements. For a given azimuth angle η, the phasing of emitted waveform from each element can easily be computed. Consider the case of equal phase increments $\Delta\phi$ between elements. For the second element from the center, we have

$$\begin{aligned} \text{Distance from axis } AA' &= 2\text{d} \sin \eta \\ &= 2\text{d} \sin \eta \frac{2\pi}{\lambda} \text{ (rad)} \end{aligned} \tag{2.77}$$

Equating this value to the phase increment at this location of $2\Delta\phi$, we get

$$\Delta\phi = \frac{2\pi d}{\lambda} \sin \eta \tag{2.78}$$

For a desired location of antenna beam η, the phase increment $\Delta\phi$ of Eq. (2.78) can be computed.

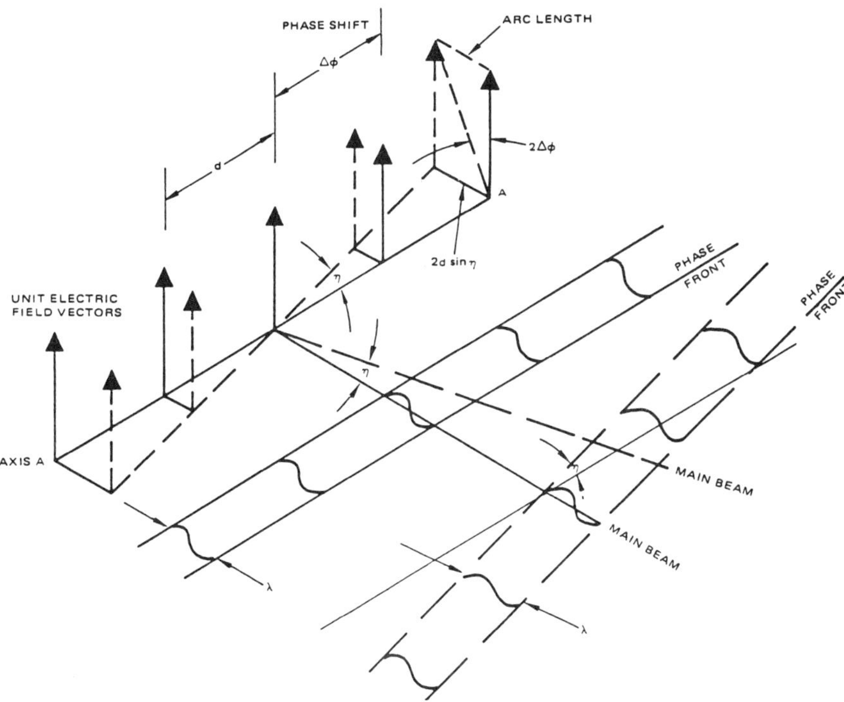

Figure 2.38 Phase fronts in electronic scanning antennas.

Equation (2.78) was derived for a one-dimensional antenna. This equation can be extended to a two-dimensional array antenna, where radiating elements are located on a flat surface and each element is equipped with a phase shifter. For a given angular beam position in azimuth and elevation, the proper phase shift of each element is computed by a digital computer. This phase shift is then put in the appropriate phase shifter for the desired spatial beam position.

In addition to eliminating the mechanical motion of the antenna, in electronic scan radar systems, the antenna beam can be moved quickly from one point to the next. This feature allows simultaneous tracking of a greater number of targets.

Figure 2.39 illustrates a ground-based, electronically scanned radar system. These systems are usually equipped with antennas having thousands of radiating elements. This figure shows a pair of phased array antennas facing 120° apart, which are designed for 240° frontal coverage (120° each). The electronically steered beam is formed from signals radiating from 1792 elements. The face of the antenna is 102 ft wide.

Figure 2.39 Electronic scan radar system (courtesy of Raytheon Company).

2.15 PROBABILITY OF DETECTION AND S/N RATIO

In the previous discussion of detection process, it was assumed that achieving a certain S/N ratio is sufficient for detection. Actually, more sophisticated methods of calculating the probability of detection are available that relate S/N ratio to threshold setting and provide the probability of target detection at a given radar scan. These calculations involve assumptions regarding the type of targets and the effective radar signal return from these targets. The threshold setting is directly related to the number of false target detections due to thermal noise of the radar receiver system, as will be seen in the following discussion.

2.15.1 Types of Radar Target Returns

Radar targets can be divided into two general categories for the purposes of *radar target cross section* (RCS) determination. Airborne targets can be represented by a finite number of scattering points that intercept the transmitted electromagnetic energy and reflect portions of it toward the transmitting antenna. Assuming that each reflecting point on the target reflects energy (during radar target illumination time) with a Gaussian distribution; the total reflected energy from the target will be in the form of Rayleigh distribution as shown in Figure 2.40 and discussed in reference [9]. Figure 2.40 shows experimentally measured samples of RCS values for a nose aspect airborne target. The average value of these measurements is 6.93 m^2. It can be shown that the distribution of Figure 2.40 can be approximated either by a Rayleigh or by an exponential probability distribution. This probability distribution of RCS is used in the probability of detection calculations. In the radar range equation, however, the average value of the RCS is used.

It can also be shown that targets with a dominating reflector will produce radar RCS values that have a Gaussian distribution with the average value closely approximating that of the dominating reflector. Figure 2.41 shows one such target, consisting of an augmentor that was used to enhance target RCS for detection and tracking purposes. Similar types of target returns are expected from satellites, which usually have a dominating spherical or cylindrical structure surrounded by peripheral equipment. The average RCS of Figure 2.41 is 82.3 m^2 with a standard deviation of 32 m^2.

Note that the discussed target variations of Figures 2.40 and 2.41 occur between radar scans of the target when the radar is in the search mode of operation. These variations are commonly referred to as target fluctuations or target scintillation.

2.15.2 Probability of Detection

These variations of target RCS values can be incorporated in the probability of detection of a sinusoidal signal in Gaussian noise [9]. The calculations result in the probability of detection *versus* S/N ratio curves of Figures 2.42 and 2.43. These curves use the probability of false alarm, P_n, as a parameter. The value of P_n is utilized in computations relating

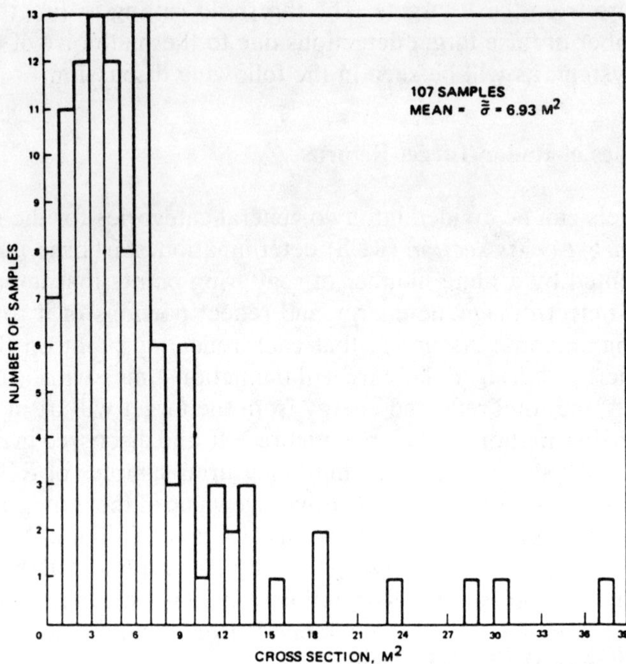

Figure 2.40 Experimentally measured target radar cross section of an aircraft or other airborne target.

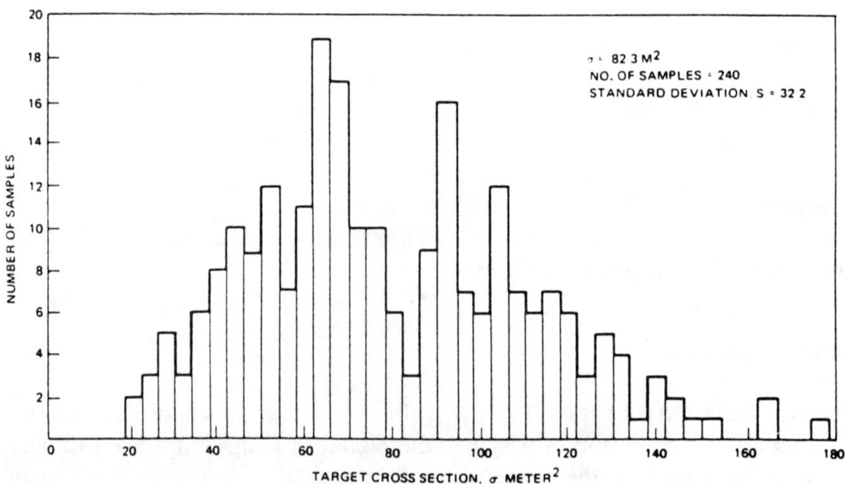

Figure 2.41 Target cross section data obtained from an augmented target.

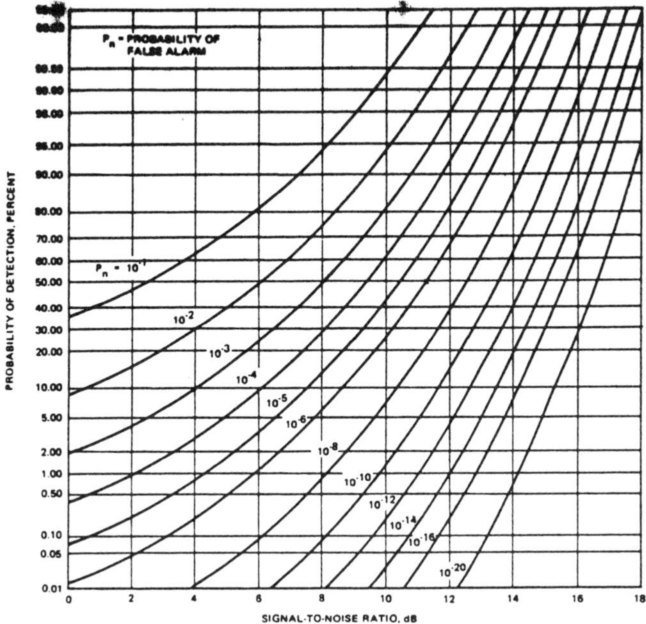

Figure 2.42 Probability of detection *versus* S/N ratio for a nonscintillating target.

to threshold setting and false alarm rates. Figure 2.42 gives the probability of detection for a nonscintillating target. This curve can also be used for calculation where the target consists of a dominating reflector.

Figure 2.43 gives the probability of detection curves for a scintillating target with exponential scintillation. As just mentioned, exponential scintillation results from electromagnetic radiation return of targets with several "equal strength" reflectors.

The threshold setting can be computed by using the 50 percent probability of detection S/N ratio of a nonscintillating target, as given in Figure 2.42. For example, a probability of false alarm P_n of 10^{-8} will require a threshold setting of 12.5 dB.

2.15.3 False Alarm Rate

The threshold setting is usually computed from false alarm requirements. The requirements are usually specified in terms of false alarms per unit time: per second or per minute. Consider a low PRF radar system with a pulse duration of τ seconds and range gates of equal duration. The number of range gates examined per unit time will be

$$n = \frac{1}{\tau} \tag{2.79}$$

where n is the number of range gates. Using P_n as the probability of false alarm in each range gate, the number of false alarms per unit time n_{fa} will be

$$n_{\text{fa}} = \frac{P_n}{\tau} \tag{2.80}$$

Or, conversely, time between false alarms T_{fa} will be

$$T_{\text{fa}} = \frac{\tau}{P_n} \tag{2.81}$$

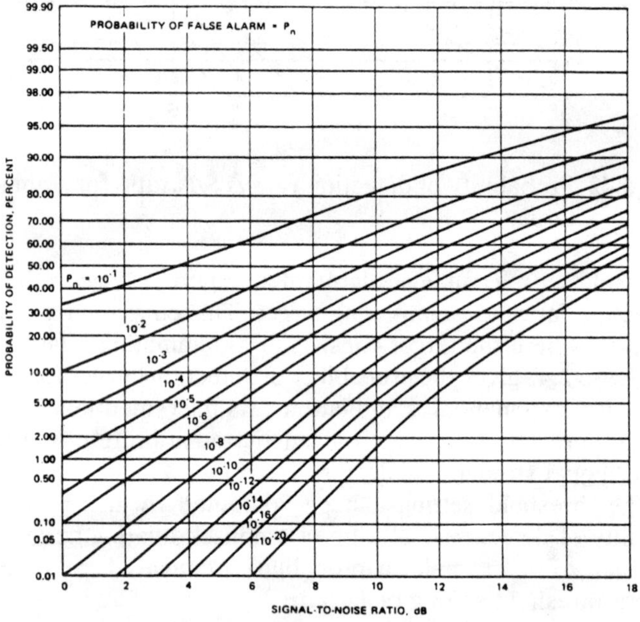

Figure 2.43 Probability of detection *versus* S/N ratio for scintillating target with exponential scintillation.

The false alarm rate of a high PRF radar can be computed by a consideration of the number of detection filters examined for the presence of target signal and the time constant of each filter, thus:

$$T_{fa} = \frac{T_c}{NP_n} \tag{2.82}$$

where T_c is the time constant of each filter and N is the number of doppler filters.

2.15.4 Numerical Example

Consider a low PRF radar system with a pulse duration of 0.1 μs. We want a time between false alarms of 10 minutes. Compute the probability of false alarm and the resulting threshold setting. From Eq. (2.81), we have

$$\begin{aligned} P_n &= \frac{\tau}{T_{fa}} \\ &= \frac{0.1E-6}{(10)(60)} \\ &= 1.66E-10 \\ &\simeq 2.0E-10 \end{aligned} \tag{2.83}$$

The resulting threshold setting from Figure 2.42 will be 13.5 dB, which corresponds to 50 percent probability of detection value with probability of false alarm of Eq. (2.83). Furthermore, assume that a target S/N ratio of 12 dB has been computed for a target radar range of 100 km. Compute the probability of detection for this S/N for nonscintillating and exponentially scintillating targets.

From Figure 2.42 for $P_n = 2E - 10$ we will obtain a probability of detection of 15 percent for a nonscintillating target and from Figure 2.43 we will obtain a probability of detection of 25 percent for an exponentially scintillating target.

In actual application of the probability of detection curves, the values of S/N are calculated as a function of radar-target range. The single look (at each radar scan) probability of detection values corresponding to each S/N ratio is obtained from Figures 2.42 or 2.43 depending on the type of target under consideration. These values will specify the probability of target detection as a function of range. Furthermore, the single look probability values can be used to compute cumulative probability of detection values as the target-radar range decreases. [1]

2.16 SUMMARY

In modern-day radar systems, the radar itself provides the sensing device of the system, while the necessary target tracking and fire control system calculations are performed by a digital computer. With advances made in radar hardware technology, such areas as coherent signal transmission, PRF switching, and multiple PRF generation no longer present a problem. Advances in digital computer technology make it possible to perform a large number of calculations in very little time. The combination of these factors enables us to switch several PRFs and make the necessary calculations to obtain target range and range rate, all within the target illumination period of about 20 ms.

In addition, present-day technology enables us to use the radar mode best suited for existing conditions, be it low, high, or medium PRF. The high PRF mode is used primarily for long-range detection and tracking of incoming targets in doppler frequency. This frequency discrimination produces accurate range rate (on the order of feet per second) values. The high PRF doppler mode, of course, results in good target detection despite ground clutter. The ability of high PRF radars to obtain accurate radar-target range is, however, limited (in order of miles). To obtain accurate range, and also for certain ground-mapping applications, we may employ the low PRF mode of the radar system. These radars result in target range accuracies in the order of feet. They also produce high resolution ground maps.

The medium PRF mode can be considered a compromise between high and low modes. It results in reasonably accurate range (on the order of thousands of feet) and range rate (on the order of hundreds of feet per second) values and acceptable capability in detecting targets in the presence of ground clutter. The medium PRF mode, however, does not enjoy the long-range capability of the high PRF radars.

An example of a radar with several modes of operation is the AN/APG-66 radar of the F-16 aircraft, as described in [3]. This radar has several modes of operation using a variety of waveform designs. Some of these programmed modes of operation are as follows:

Downlook—a coherent medium PRF, pulsed doppler mode used for detecting low-flying airborne targets in the presence of mainlobe clutter.

Uplook—a coherent, low PRF mode used in the absence of clutter to detect airborne targets.

Real beam ground map—a noncoherent, frequency-agile mode for air-to-surface mapping used to identify ground targets.

Doppler beam sharpening—a ground mapping mode that employs doppler processing to enhance mapping resolution.
Ship detection—ocean surveillance mode for detection of ships in the presence of sea clutter.

These modes of operation require great variation of the transmitted waveforms, as well as extensive signal processing of the target returned signals. Mechanized characteristics of this radar are as follows:

Volume	= 4 ft^3
Weight	= 300 lb
Power	= 3.7 kW
Cooling	= air cooled at 12 lb/min
Electronic parts	= 9500
Reliability	= 97-h demonstrated MTBF
Maintenance	= 5-min flightline *mean time to repair* (MTTR)

REFERENCES

1. S. A. Hovanessian, *Radar System Design and Analysis*, Artech House, Dedham, MA, 1984.
2. S. A. Hovanessian, "An Algorithm for the Calculation of Range in Multiple PRF Radars," *IEEE Trans. on AES*, March 1976.
3. W. H. Long and K. A. Harriger, "Medium PRF for the AN/APG-66 Radar," *Proc. IEEE*, Feburary 1985.
4. B. L. Lewis, F. F. Kretschmer, Jr., and W. Shelton, *Aspects of Radar Signal Processing*, Artech House, Norwood, MA, 1986.
5. S. S. Blackman, *Multiple-Target Tracking with Radar Applications*, Artech House, Norwood, MA, 1986.
6. A. I. Leonov and K. I. Fomichev, *Monopulse Radar*, Artech House, Norwood, MA, 1986.
7. T. J. Nessmith, "Range Instrumentation Radars," *IEEE Trans. AES*, November 1976.
8. E. Brookner, "Phased Array Radars," *Scientific American*, February 1985, pp. 94–99.
9. S. A. Hovanessian, H. H. Ahn, "Calculation of Probability Detection with Target Scintillation," IEEE Trans. *AES,* March 1973, p. 300.

Chapter 3
Synthetic Array and Imaging Radars

Airborne and spaceborne microwave radars are currently used to make images of the ground below for aerial mapping, earth exploration, navigation, and target identification and designation. The microwave radar images obtained at centimeter wavelengths complement the thermal imager pictures at μm wavelengths in studies involving ground terrain and earth resources.

This chapter discusses the fundamentals of radar mapping and imaging. The x and y direction resolution characteristics and limitations as well as S/N ratio calculations and power requirements are discussed. Several examples of airborne and spaceborne imagery illustrate the covered principles.

3.1 INTRODUCTION

The first aerial images of the earth were crude photographs taken from balloons in the middle of the nineteenth century. This initial step was followed by aerial photography, which takes place in the visible wavelength region of 0.4–0.8 μm, infrared imagery in the range of 1–15 μm, and microwave radar imagery in the range of 1–30 cm.

Radar sensors are capable of providing the high-quality real-time imagery required for navigation, aerial mapping, and target identification and designation, unimpeded by weather conditions, whereas passive devices operating in the visual and infrared regions are degraded by atmospheric conditions. The atmosphere absorbs a significant fraction of the emitted radiation in the visible and infrared regions, even on clear days. In overcast or rainy weather, the performance of visible and infrared detectors is impaired, because moisture absorbs greater quantities of the emitted energy. To overcome dependence on weather and atmospheric conditions, an imaging device must provide its own source of illumination. Additionally, the transmitted energy from this source must not incur a great loss while traveling through the atmosphere.

Radar systems operating at wavelengths of longer than 1 cm fulfill both requirements. Radars provide their own source of illumination by transmitting electromagnetic energy. At wavelengths of greater than 1 cm, atmospheric absorption of this energy is minimal. Therefore, radar imaging systems essentially operate under all weather conditions. Imaging systems operating at wavelengths on the order of centimeters cannot provide the same detailed image resolution obtainable in the visible region. However, radar wavelengths are short enough to obtain remarkably accurate imagery for geographical, geological, and military purposes. The key features of radar mapping are summarized in Table 3.1.

Table 3.1 Key Features of Radar Mapping

- Provides its own source of illumination
- All weather, day or night operation
- Large area coverage in short time
- Additional target features derived from use of longer wavelengths are complementary to data from other sources
- Surface features displayed despite vegetation and cloud cover
- Storage and transmission of raw data
- Operates at relatively low grazing angle
- Presents a plan view image of offset area

In addition to providing its own source of target illumination, radar has several other advantages. Because of the altitude of the vehicle bearing the mapping radar, large surface areas can be covered. Radar wavelengths are complementary to other mapping wavelengths in the visible and infrared regions. Radar maps, photographs, and infrared maps can be studied simultaneously to obtain additional information about the characteristics of the surface terrain or manufactured objects.

Another important feature of radar mapping (especially with digital signal processing) is that data can be electronically stored and transmitted. This is particularly important in tactical operations, because target information can be relayed electronically and maps of the area made remotely.

3.2 IMAGING RADARS

There are two broad categories of imaging radars: real array, and synthetic array [1 and 2], or synthetic aperture radar. In antenna theory, however, the word *array* is used for antennas in which individual radiating elements are separate rather than part of a continuous radiator, while *aperture* designates a class of continuous radiators over a given surface.

Using the coordinate system of Figure 3.1, both types, real and synthetic, obtain x direction (or range resolution) by controlling the transmitted pulsewidth or by suitably encoding it to permit pulse compression upon reception. Real array radars obtain the y direction (or azimuth resolution) by the antenna beamwidth. Since the antenna beamwidth is inversely proportional to antenna length, the y direction resolution of real array imaging radars can be improved by using long antennas (narrow beamwidth). To obtain y direction resolution comparable to that of the x direction, the length of the antenna is such that it can be placed only along the belly of the aircraft looking to one side; hence, the name *side-looking radars*.

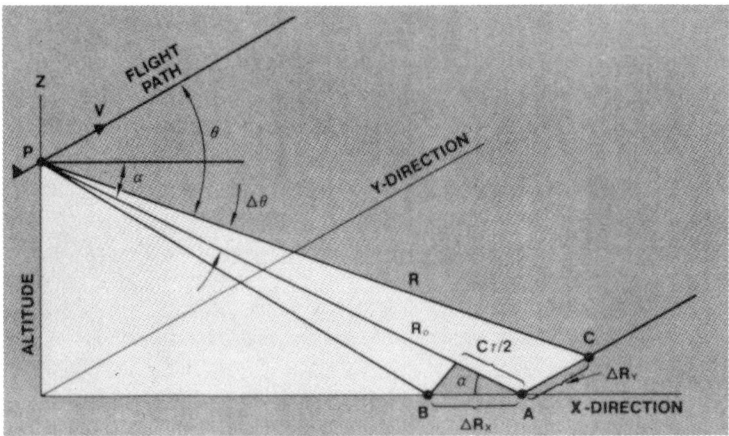

Figure 3.1 Flight path direction and x and y direction resolution.

The desire to use smaller airborne antennas while increasing the y direction resolution for ground mapping led to the development of *synthetic array radars* (SAR). Like real array imaging radars, these use transmitted pulsewidth for x direction resolution. However, synthetic array radars use the incremental doppler shift between adjacent points on the ground, rather than the antenna beamwidth, for y direction resolution.

The processing of this doppler shift effectively converts the wide beam of a small antenna to a series of narrow beams, giving better resolution. The development of radar systems using SAR began with *doppler beam-sharpening* (DBS), which was invented in 1951 by Carl Wiley at Goodyear Aircraft Company.

It is recognized that the definitions of doppler beam-sharpening and synthetic array radars are not standard within the technical community. The following definitions are used in this article: in DBS, the radar beam is divided into a number of subbeams, with each subbeam equivalent to a resolution cell, as shown in Figure 3.2. The velocity component (or doppler shift) along each subbeam is used to obtain cross-range resolution. For this reason, in the DBS method of ground mapping, resolution increases as range decreases. In synthetic array radar, the processing technique keeps the actual resolution constant with range.

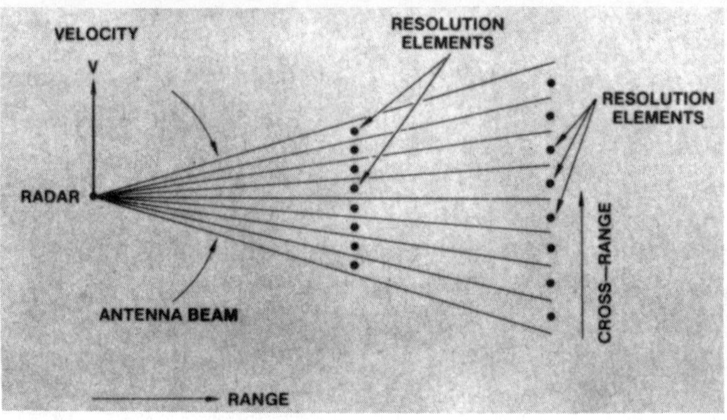

Figure 3.2 Doppler beam sharpening of the antenna beam.

Figure 3.3 shows ground maps obtained by real array and DBS airborne radars using the same size antenna. Note the remarkable quality of the doppler ground map compared with the real array map. Another important observation concerning Figure 3.3 is the orientation of these ground maps. The real array map represents the area directly ahead of the aircraft. The doppler ground map represents the area to the side of the aircraft, as only the resolution cells to the side can be separated by their doppler shifts.

Figure 3.4 illustrates sophisticated SAR imagery. It is an image of Los Angeles, California, produced from the data of SEASAT satellite, obtained on August 12, 1978. The data were obtained by the Jet Propulsion Laboratory of the California Institute of Technology and digitally processed by MacDonald Dettwiler and Associates of Richmond, British Columbia, Canada. The resolution of this image is 25 m, and 32 shades of gray were used to produce this picture.

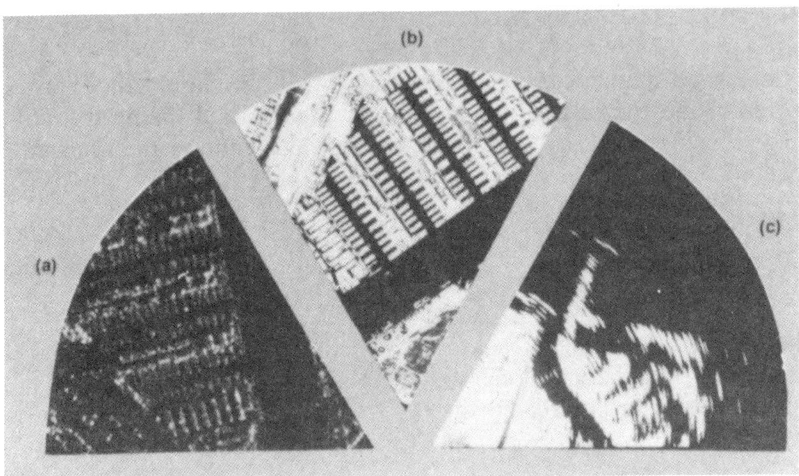

Figure 3.3 Doppler ground map with 20 ft resolution (a), a photograph of the same area (b), and a real beam map with 3° antenna beam and 35 ft range resolution (c).

Figure 3.4 SAR image of Los Angeles, California, obtained from satellite data.

3.3 *X* DIRECTION RESOLUTION

Both real and synthetic array imaging radars measure the distance between the radar and the target by transmitting a pulse and timing its return. Range resolution ΔR is determined by the pulsewidth of the transmitted pulse as

$$\Delta R = \frac{c\tau}{2} \qquad (3.1)$$

where ΔR is the range resolution, τ is the pulse width, and c is the velocity of propagation of electromagnetic energy.

Referring to Figure 3.1, the resolution along the x axis becomes

$$\Delta R_x = \frac{c\tau}{2 \cos \alpha} \qquad (3.2)$$

where α is the depression angle in the ZX plane. For a pulsewidth of 1 μs, Eq. (3.2) results in an x direction resolution of about 150 m. This resolution can be improved easily by using either analog (frequency modulation) or digital (Barker code) pulse compression methods. In these methods, the transmitted pulse is coded and, upon reception, the pulse is passed through a decoding matched filter. The frequency-time characteristics of the transmitted pulse is shown in Figure 3.5. The pulse of duration τ is frequency modulated from f_1 to f_2 at a rate of Δ/τ where $\Delta = f_2 - f_1$. The effect of frequency modulation on the transmitted sinusoidal signal is shown in Figure 3.5(c).

The target return signal at the receiver will be similar to the transmitted signal in frequency-time characteristics, as shown in Figures 3.6(a) and 3.6(b). The receiver, however, acts on the received signal by a time delay network with delay-frequency characteristics as given in Figure 3.6(c). From this figure it can be seen that the lowest received frequency, f_1, is delayed the longest, while the highest received frequency, f_2, is not delayed at all. In this manner, the initially received frequencies are made to "wait" while subsequently received higher frequencies are delayed by shorter amounts.

The final output, by conservation of energy, is a pulse of short duration, τ', and large amplitude, as shown in Figure 3.6(d). This pulse will have a (sin x/x) characteristic with a maximum value of

$$\sqrt{D}$$

where D is defined as the dispersion factor and is equal to $\tau\Delta$ where Δ is the frequency excursion. The pulse compression, PC, ratio is given by the ratio of transmitted and compressed pulses:

$$\mathrm{PC} = \frac{\tau}{\tau'}$$

This value determines the achieved resolution of the system. For example, a 1 μs pulse with a PC of 100 : 1 will result in an approximate x direction resolution of 1.5 m.

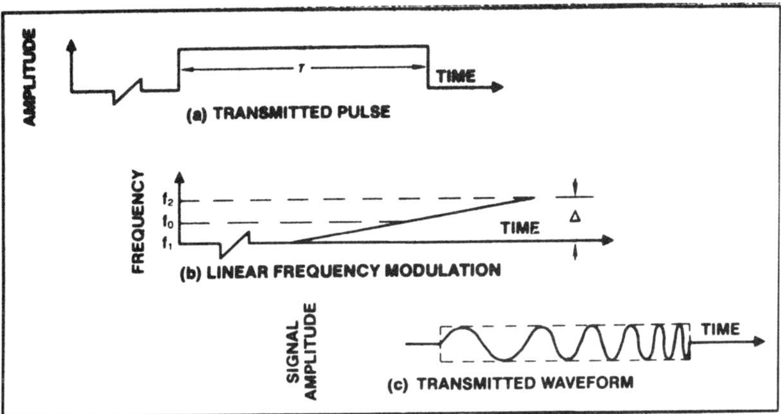

Figure 3.5 Transmitted waveform of a linear FM pulse.

3.3.1 Digital Pulse Compression

Digital codes can also be used to improve x direction resolution in SAR radars. A prime example of these codes is *binary phase code* (BPC). A binary phase code of compression ratio, $\xi = T\Delta$, where T and Δ are defined as the pulse length and the signal bandwidth, respectively, consists of a time sequence of ξ signals each of duration $1/\Delta$, at a constant carrier frequency. Each segment is characterized by being in phase (+1) or 180° out of phase (−1) relative to some *continuous wave* (CW) signal. Hence, a BPC can be represented by a sequence of ξ symbols, each plus or minus.

A particular class of BPC known as Barker codes have quite low sidelobes. Let us consider the longest known Barker code, which contains

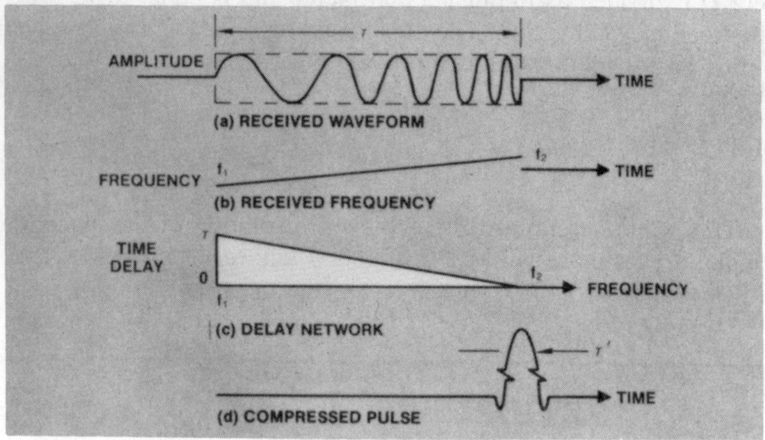

Figure 3.6 Received waveform of the FM pulse and subsequent pulse compression.

13 elements. This is illustrated in Figure 3.7, which shows the binary sequence, the corresponding video signal as a frequency-modulated squarewave, and the phase reversals of the transmitted CW signal. If such a sequence is transmitted and received, a suitable decoder consists of a tapped delay line, shown in Figure 3.8, in which each of the 13 taps adds or subtracts according to the code. The total delay is T and the individual delays are $1/\Delta$, the length of a segment. Figure 3.9 gives the procedure for calculating the decoder output, and Figure 3.10 gives a plot of this output. The input signal appears in the decoder for 26 time segments from first appearance at tap 1 to last appearance at tap 13.

Only the Barker codes of 13 elements or less have the unique property of unit sidelobes. However, many longer codes have peak values in the hundreds and thousands (References [3] and [4]) with larger peak sidelobes. None of these codes have sidelobes of $1/N$ times the peak value or less, where N is code length, as was just demonstrated for the Barker code of length 13.

3.4 *Y* DIRECTION RESOLUTION

For a given wavelength and antenna size the y direction (azimuth direction) resolution of a real array imaging radar, looking sideways, is proportional to range

$$\Delta R_y = \frac{R\lambda}{l} \tag{3.4}$$

where R is the range, λ is the wavelength, and l is the antenna length. In Eq. (3.4) the quantity λ/l represents the antenna beamwidth in radians. For example, a side-looking radar with a wavelength of 1.5 cm, operating at 5 km from the target and having a length of 5 m, yields a y direction resolution of 15 m. Using the same parameters, a tactical fighter antenna of 0.75-m diameter will give a resolution of 100 m. This resolution, although sufficient to detect broad outlines of large areas, is inadequate against tactical targets.

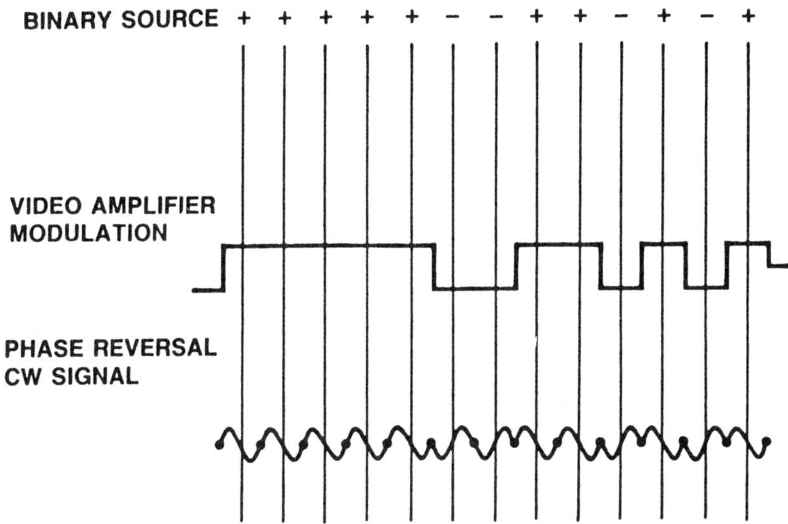

Figure 3.7 Waveforms for a Barker code 13 elements long.

In synthetic array radars, resolution in the y direction is determined by the frequency shift produced by the doppler phenomenon. Consider Figure 3.1, the doppler shift along R will be

$$f = \frac{2V\cos\theta}{\lambda} \tag{3.5}$$

where f is the doppler shift, V is the velocity, λ is the wavelength, and Θ is the subtended angle between the velocity vector and R.

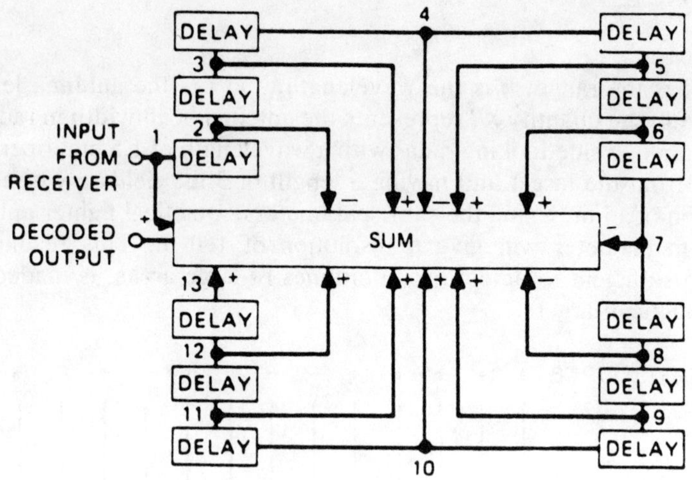

Figure 3.8 Barker decoder 13 elements long.

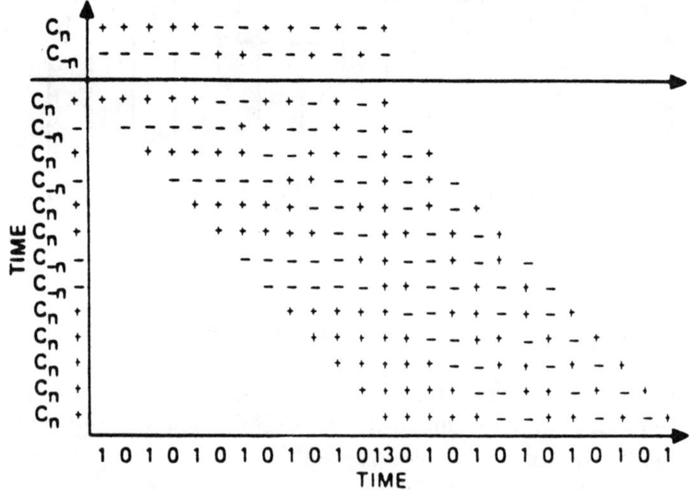

Figure 3.9 Procedure for calculating the decoder output for a Barker code containing 13 elements.

The rate of change of this shift, with respect to θ, can be obtained by differentiation (neglecting signs):

$$\Delta f = \frac{2V \sin \theta \Delta\theta}{\lambda} \tag{3.6}$$

where difference notation is used. The value of $\Delta\theta$ can be obtained in terms of range R and y direction resolution ΔR_y:

$$\Delta\theta = \frac{\Delta R_y}{R} \tag{3.7}$$

Using Eq. (3.7) in Eq. (3.6) and solving for ΔR_y, we have

$$\Delta R_y = \frac{R\lambda}{2V \sin \theta} \Delta f \tag{3.8}$$

Equation (3.8) gives the y direction resolution in terms of range, wavelength, aircraft velocity, angle, and most important, frequency difference between points A and C of the figure. Thus, the calculation of the y direction resolution has been reduced to a frequency measurement. Assuming we can measure doppler shifts within 10 Hz, and using the parameters R = 10 km, V = 200 m/s, λ = 3 cm, θ = 90°, Equation (3.8) results in ΔR_y = 7.5 m.

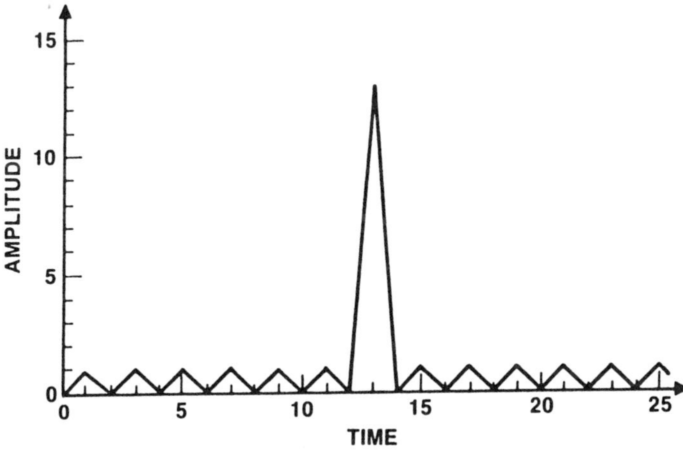

Figure 3.10 Decoded Barker code 13 elements long.

From the previous discussions, it can be seen that SAR radars produce ground maps of significantly better quality than real beam radars, as illustrated in Figure 3.3. Synthetic array radars, however, cannot map areas directly ahead of the aircraft, because resolution cells directly ahead of the aircraft produce minimal doppler shift separation, due to the small subtended angles between those cells and the aircraft velocity vector. Areas directly in front of the aircraft can be mapped by real beam radars and filled-in DBS or SAR maps.

3.5 SIGNAL PROCESSING OF SAR DATA

3.5.1 Doppler Signal Processing

In synthetic array radars, signal processing consists of collecting and processing radar data to produce radar imagery. This data is processed across all range resolution cells to produce x direction resolution. Then each range resolution cell data is doppler processed to produce azimuth or y direction resolution. This process produces a two-dimensional image of the mapped area, which can be displayed in real time. Figure 3.11 gives a description of signal processing procedure. Range Gates 1 through N represent x direction resolution cells. The processing of the data of each resolution cell, which can be done either through doppler filters or digital spectrum analysis, will produce the contents of the y direction resolution cells. Note that the contents of each resolution cell is in an S/N ratio, which is converted to shades of gray in producing the displayed imagery.

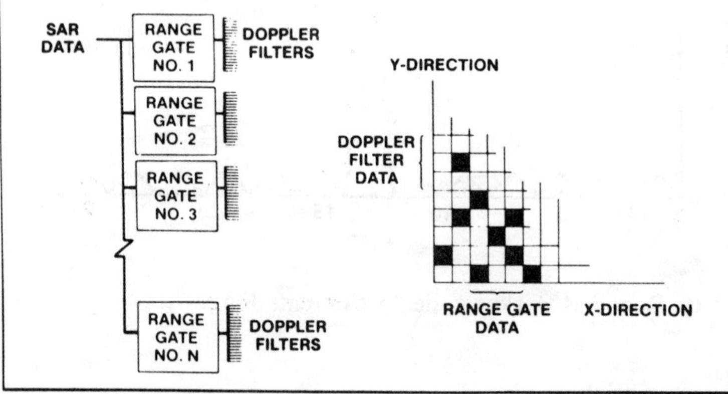

Figure 3.11 SAR data processing through range gates and doppler filters.

To gain an understanding of the signal processing in the y direction, consider a single point on the ground at range R_0. As the radar moves at a velocity, V, perpendicular to this point (Figure 3.1), the radar-ground point range as a function of time can be written

$$R = R_0 + \frac{1}{2}at^2 \tag{3.9}$$

where a is the acceleration along R_0, and t is the time. Using centripetal acceleration V^2/R for a, we get

$$R - R_0 = \frac{1}{2}\frac{V^2}{R}t^2 \tag{3.10}$$

The range differential $R - R_0$ can be converted to phase differential by noting that each 2π rad in phase is equal to one wavelength, λ. Using this, the one-way phase differential as a function of time t will equal to $\Delta\phi$:

$$\Delta\phi = \frac{\pi V^2 t^2}{\lambda R} \tag{3.11}$$

With the two-way phase differential equaling twice this value:

$$\Delta\phi = \frac{2\pi V^2 t^2}{\lambda R} \tag{3.12}$$

Figure 3.12 gives the quadratic phase history of a single ground point as derived in Eq. (3.12). The upper right-hand section of the figure gives the same history after subtracting 2π rad and implementing a periodic reference signal in every 2π rad slot. The bottom of the figure shows the resulting zone plate of white and black regions after thresholding. Optical processing methods primarily involve processing the "zone plate" data from each ground point. This is accomplished by using coherent reference light and holographic methods [5].

Figure 3.13 shows flight geometry and radar recording of synthetic array data for two near and far ground points covered by the antenna beam. The same figure also shows an optical computer for processing this data. The conical lens of the figure is used to take out differences in the recorded data due to radar target range, making the light rays parallel. The balance of the lens system shown brings the light produced by conical lens into focus in the output plane as two distinct ground points.

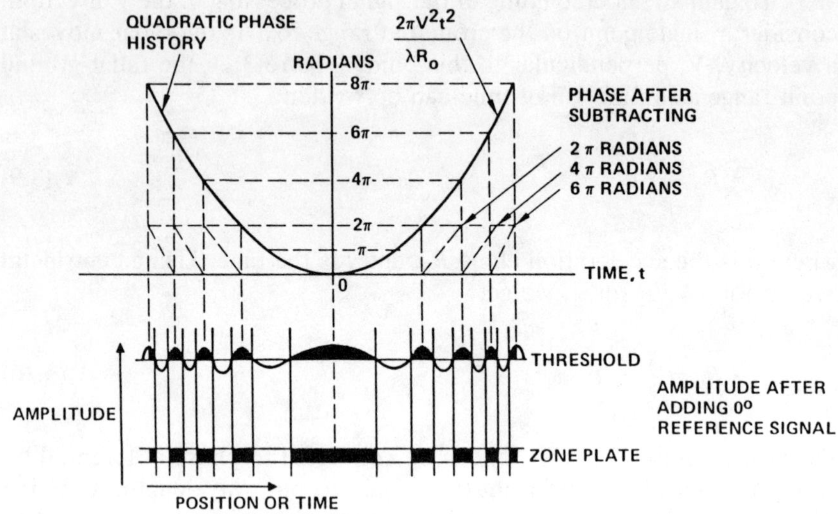

Figure 3.12 Quadratic phase history of a single ground point, subtraction of complete cycles, and production of zone plate.

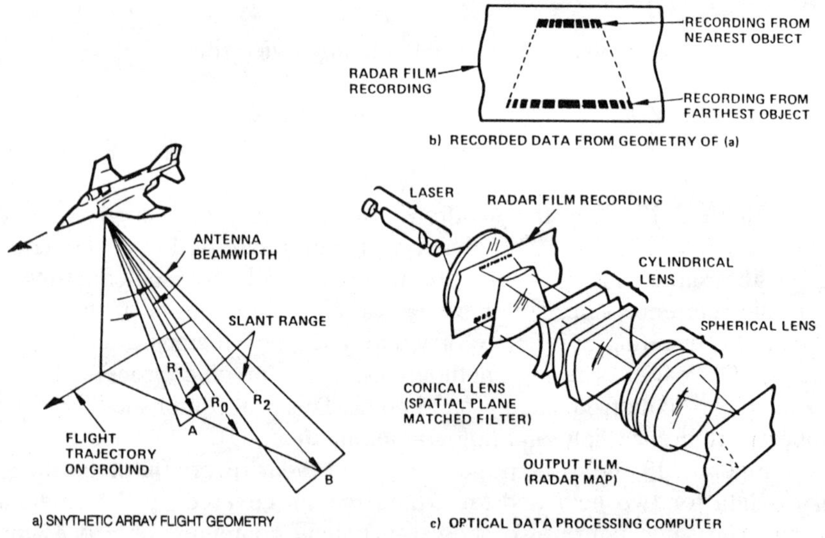

Figure 3.13 Optical processing of synthetic array radar data.

3.5.2 Digital Signal Processing

Consider the quadratic phase history of a single ground point as given in Figure 3.12. Furthermore, observe that the derivative of phase with respect to time represents a frequency change:

$$2\pi f = \frac{d\phi}{dt} \tag{3.13}$$

Using the phase change relation of Eq. (3.12) in Eq. (3.13), we get

$$f = \frac{2V^2t}{\lambda R} \tag{3.14}$$

where f represents the frequency variation of a single ground point as a function of time. Figure 3.14 shows both the phase and frequency variations of a single ground point. Note the linear frequency-time relationship. Note also the center of the figure is at zero time with negative and positive axes.

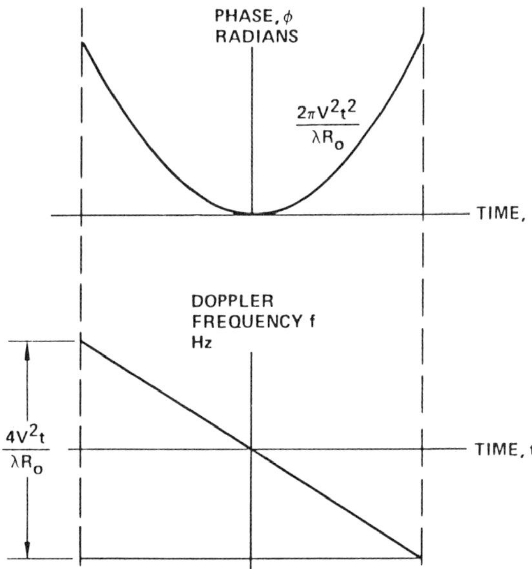

Figure 3.14 Quadratic phase and linear frequency signal history of a single ground point.

With this background discussion of a single point signal return, we will now consider the case of several illuminated ground points, as shown in Figure 3.15. In this figure assume that the aircraft flies with a velocity, V, and its beamwidth can illuminate six points on the ground. In position A, the radar begins to illuminate point 1; and at position B, it completes the illumination of this point. At position A, the component of aircraft velocity along the vector $A-1$ will give rise to the frequency shift in the frequency-time diagram. As the aircraft flies from A to B this frequency will decrease, passing through zero as shown in the bottom of the figure. Other points of the figure will also have the same frequency history, with the total frequency excursion on both sides of zero frequency equaling $4V^2t/\lambda R_0$, twice the value given by Eq. (3.14).

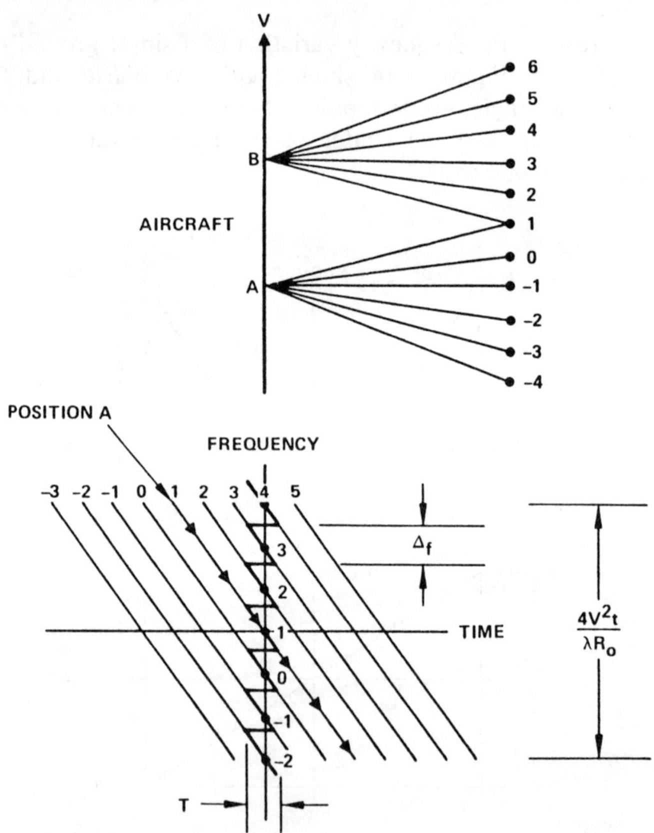

Figure 3.15 Frequency-time signal history of several ground points.

Digital signal processing of SAR data involves utilizing the frequency-time signal of Figure 3.15, sorting this data, and assigning it to the appropriate resolution cells. There are two methods generally available to implement this procedure.

One method uses matched filters similar to the pulse compression filters of Figures 3.5 and 3.6. Each filter has time-frequency characteristics to match one of the slant frequency lines, starting at points -3, -2, -1, . . . , of Figure 3.15. Each filter, in effect, follows one of these lines and puts out a burst of power at the end corresponding to a specific resolution cell.

The second method uses a slanted time-frequency reference signal that, in effect, rotates the slanted frequency excursions of Figure 3.15 so that they parallel the time axis of the figure. Each frequency line then intercepts the frequency axis at designated frequencies, -2, -1, 0, 1, 2, ...

This operation is implemented through a swept oscillator with the same slope as the target return signal. The target return of various ground points is mixed with the swept oscillator signal producing a constant frequency signal for each target return.

3.6 RESOLUTION LIMITS OF SAR SYSTEMS

The x direction or range resolution of SAR systems is determined by the processed pulsewidth, τ. This resolution is given by

$$\Delta R_x = \frac{c\tau}{2} = \frac{c}{2\Delta} \tag{3.15}$$

where τ is the pulsewidth and Δ is the compression bandwidth, as discussed in Figures 3.5 and 3.6.

The y direction resolution, as discussed earlier, is determined by doppler processing the data. This processing can take place by using various time durations of ground point signal returns.

If we use the nonoverlapping portion of the data, which eliminates the need for transverse (time dependent) filters, we will be dealing with the duration of data designated by T of Figure 3.15. This is the so-called *unfocused* case. The corresponding frequency excursion in this case will be

$$\Delta f = \frac{2V^2T}{\lambda R_0} \tag{3.16}$$

Equating the bandwidth Δf to $1/T$, we get

$$\begin{aligned} VT &= \sqrt{\lambda R/2} \\ &= 0.707\sqrt{\lambda R} \end{aligned} \tag{3.17}$$

The quantity VT represents the time to fly-out one filter time constant, which is also equal to the resolution in the y direction. Thus, for the *unfocused* case we have

$$\Delta R_y = 0.707\sqrt{\lambda R} \tag{3.18}$$

From Eq. (3.18) the resolution in the unfocused case is proportional to the square root of range R, as compared to the first power of R in the case of real array radar of Eq. (3.4).

The limit of the y direction resolution, where longer lengths of data are used, the so-called *focused* case can be obtained from the following discussion. The longest length for which data is available to the radar system is a function of the antenna coverage on the ground. This ground coverage is given as

$$\begin{aligned} \Delta R_G &= R\vartheta_{BW} \\ &= \frac{R\lambda}{l} \end{aligned} \tag{3.19}$$

where ϑ_{BW} is the antenna beamwidth approximated by λ/l with λ the wavelength and l the antenna length. The time to fly-out this range can be obtained by dividing Eq. (3.19) by the radar velocity:

$$t = \frac{R\lambda}{Vl} \tag{3.20}$$

where t is the time and V is the radar velocity. For this duration of time, the frequency excursion from Eq. (3.16) will be

$$\Delta f = \frac{2V}{l} \tag{3.21}$$

The equivalent time constant of the Δf frequency excursion is $1/\tau'$; using this in Eq. (3.21) we get

$$V\tau' = l/2 \tag{3.22}$$

Denoting $V\tau'$, the distance traveled in one filter time constant, as resolution in the y or azimuth direction, we have the limit of this resolution from Eq. (3.22) as

$$\Delta R_y = l/2 \tag{3.23}$$

That is, the y direction resolution is limited to one-half of the antenna length.

3.7 S/N RATIO AND POWER REQUIREMENTS

The SAR system power requirements can be computed by using the matched transmit-receive radar equation and substituting the ground resolution cell equivalent cross section for the radar target cross section in this equation. The radar signal to noise ratio equation is given as

$$\text{S/N} = \frac{\overline{P}T_i G^2 \lambda^2 \sigma}{(4\pi)^3 R^4 (kT) L} \tag{3.24}$$

where

S/N = signal-to-noise power ratio
$\overline{P}$ = average power
= $P\tau$ PRF
P = peak power
τ = pulsewidth
PRF = pulse repetition frequency
T_i = on-target time
G = antenna gain
λ = wavelength
σ = radar target cross section
R = radar target range
k = Boltzmann's constant
T = equivalent receiver noise temperature
L = transmit-receive losses

The radar target cross section of a ground resolution element can be written as

$$\sigma = (\Delta R_x)(\Delta R_y)\,\sigma_0 \tag{3.25}$$

where ΔR_x and ΔR_y are x and y direction resolution cell sizes and σ_0 is the appropriate ground backscattering coefficient. The value of σ_0 is primarily a function of surface roughness as compared to the incident wavelength. It may vary between 0.001 and 1 depending on the angle of incident, wavelength, antenna polarization, and so on.

Using the value of ΔR_y from Eq. (3.8) and noting that the on-target time, T_i, should be matched to the inverse of doppler shift Δf, we have

$$\sigma = (\Delta R_x)\left(\frac{R\lambda}{2V\sin\theta}\right)\Delta f\sigma_0$$
$$= (\Delta R_x)\left(\frac{R\lambda}{2V\sin\theta}\right)\frac{\sigma_0}{T_i} \tag{3.26}$$

Using Eq. (3.26) in Eq. (3.24) and solving for average power, $\overline{P}$, we have

$$\overline{P} = \frac{(4\pi)^3 R^3 kTL}{G^2\lambda^2\sigma_0\Delta R_x}\left(\frac{2V}{\lambda}\right)\left(\frac{\mathrm{S}}{\mathrm{N}}\right)\sin\theta \tag{3.27}$$

This equation gives the average power requirements of a SAR radar. For a given resolution cell size and ground terrain condition, the lower limit of S/N can be specified. Using this and appropriate radar parameters, average power requirement $\overline{P}$ can be computed.

The following should also be observed with respect to equation (3.27):

1. The power is proportional to R^3 instead of R^4 as in the radar equation.
2. Power is inversely proportional to ΔR_x.
3. Power is directly proportional to velocity, V.
4. Power is independent of y direction resolution ΔR_y.

3.8 MECHANIZATION BLOCK DIAGRAM

Figure 3.16 gives the mechanization block diagram of a SAR using pulse compression techniques. The stable local oscillator provides the basic frequency to the transmitter and acts as the reference frequency in determining the IF system frequency. The transmit-receive switch changes the radar function between transmission and reception of radar signals. Motion sensors measure motion of the antenna as input to a motion compensation computer for subsequent phase correction and focusing calculations. The coherent detectors take frequency information from

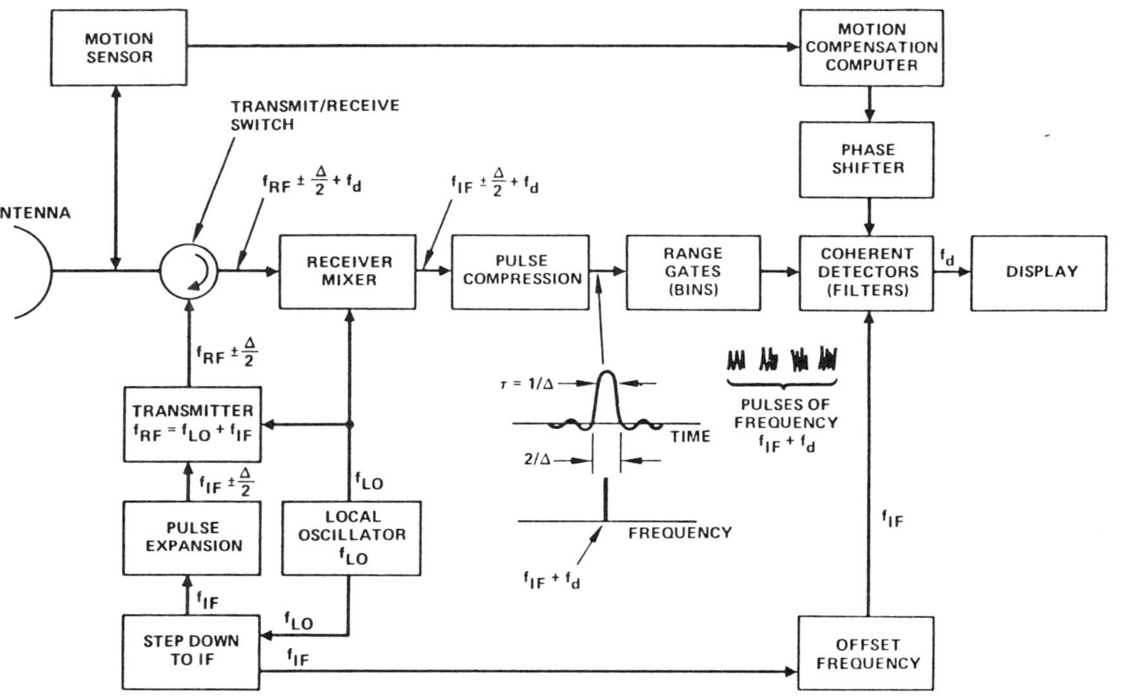

Figure 3.16 SAR mechanization block diagram.

each range bin and extract the corresponding doppler or azimuth resolution information. This filtering process is usually accomplished by a programmable digital computer implementing Fourier spectral analysis. At this point, both x and y direction resolution cells are determined and the radar image can be formed by converting the S/N ratio contents of each cell into a shade of gray.

3.9 SUMMARY

Introduction of SAR technology in a radar system means that a new capability, with high-resolution ground imaging, has been added to a familiar sensor system. Unlike the case of a real beam radar, resolution can be fixed in the design of a synthetic array radar and will not vary with radar ground mapping range. The image produced by the radar is always in plan view, regardless of grazing angle. This provides long-range capability with no reduction in mapping resolution.

When moderate shadowing takes place, the effect in general tends to be an increase in the quality of the image, since target reflectivity emphasizes the vertical dimension.

The doppler shift of the ground signal returned from about 7.5° on either side of the velocity vector is currently too small for processing, but this area can, if required, be filled in with a real beam ground map to present a more complete picture of the surface.

Spaceborne SAR systems, such as those deployed in SEASAT in 1978 and Shuttle Systems in 1984, have increased our understanding of the radar signatures of land features. The radar's ability to penetrate dry surface covers and the strong sensitivity of the radar backscattered energy to changes in the surface slope or roughness have uncovered buried river channels in southwestern Egypt. These channels, buried by 1–3 m of sand, are not visible on optical images or from the ground [7].

According to a recent article [8] remote sensing, which includes SAR technology, is the third most important commercialized part of the space industry, after communication satellites and launch vehicles. In addition to the United States and France, other countries are finding commercial uses for remote sensing sensor systems. Soyuzkarta, the Russian map making agency, is already selling high resolution images in the West. The Japanese are using an ocean sensing satellite called MOS 1, and the Canadians are planning to launch a satellite called Radarsat in the 1990s. Satellite data are used routinely for crop surveys, forestry, urban planning, oil and gas exploration, mineralogy, ocean research, and many other applications [9, 10].

Remote sensing satellites generally do not send images in the conventional sense but streams of digital data for each operating spectral band. These spectral bands may be microwave, infrared, or visible. The digital data is processed, filtered, enhanced, and otherwise developed to bring out the information desired by the customer. The last step of this process is the production of the picture or map.

Figure 3.17 shows a SAR map of Seattle, Washington. In all SAR pictures "rough" ground areas (as compared to the radar wavelength) result in large amounts of backscattered radar energy, high S/N ratios, and show up in the picture as bright spots. These include buildings and developed areas. In contrast, smooth surfaces, such as roads, still water, and aircraft runways, result in small amounts of backscattered radar energy and show up as dark regions in radar maps. Metallic structures, such as bridges, factories, refineries, and so on, because of their high degree of conductivity, result in high radar backscattering and are clearly distinguishable in radar pictures, as shown in Figure 3.17.

Figure 3.17 Synthetic array radar map of Seattle, Washington (from [8]).

REFERENCES

1. S. A. Hovanessian, *Introduction to Synthetic Array and Imaging Radars*, Artech House, Dedham, MA, 1980.
2. S. A. Hovanessian and J. C. Naviaux, "Tactical Uses of Imaging Radars," *Microwave J.*, February 1984.
3. C. E. Cook and M. Bernfeld, *Radar Signals*, Academic Press, New York, 1967, pp. 241–251.
4. R. C. Dixon, *Spread Spectrum Systems*, John Wiley and Sons, New York, 1984, p. 86.
5. W. E. Kock, *Engineering Applications of Laser and Holography*, Plenum Press, New York, 1975, pp. 196–223.
6. S. A. Hovanessian, *Radar System Design and Analysis*, Artech House, Dedham, MA, 1984, p. 96.
7. J. P. Ford, J. B. Cimino, B. Holt, and M. R. Ruzek, *Shuttle Imaging Radar Views the Earth from Challenger*, JPL Publication 86-10, March 15, 1986.
8. "Remote Sensing: New Applications Gain Acceptance," *Aviation Week and Space Tech.*, February 15, 1988, pp. 63–64.
9. C. Elachi , *Spaceborne Radar Remote Sensing,* IEEE Press, 1988.
10. J. B. Cimino, B. Holt, A. H. Richardson, *The Shuttle Imaging Radar B (SIR-B) Experiment Report,* JPL Publication 88-2, March 15, 1988.
11. F. T. Ulaby, R. K. Moore, A. K. Fung, *Microwave Remote Sensing,* Volume II, Artech House, Dedham, MA, 1982.
12. D. R. Wehner, *High Resolution Radar,* Artech House, Norwood, MA, 1987.

Chapter 4
Millimeter-Wave Radar Systems

Millimeter-wave (MMW) radar systems for ground applications are constructed at frequencies of 35, 95, 140, and 220 GHz because of the presence of *low loss* atmospheric attenuation characteristics at these frequencies. The main attraction of MMW radars lies in their smaller beamwidth for a given size antenna. This small beamwidth results in good angular resolution as well as the possibility of detecting nonmoving targets in the presence of ground clutter. In fact, in recent years, detection of nonmoving targets by airborne MMW radars mounted on helicopters have received significant technical attention, [1], [2], [3]. These nonmoving targets might be stationary ground targets, such as tanks and trucks, or airborne hovering helicopters, and the detection could take place day or night, in the presence of ground clutter return, and under adverse weather conditions, such as fog and rain. Since the detection ranges are not very long (on the order 5 km), MMW radars are well suited for this application: their higher frequencies (i.e., smaller beamwidths) allow them to subtend smaller ground clutter areas in which the target is located. This gives a better target-to-clutter-signal ratio.

A new application of MMW radars is to enhance the safety of helicopter operation by detection of hazardous cables and other small radar cross section objects during low visibility weather conditions or darkness. Microwave and millimeter-wave radars, because of their large beamwidth as compared to laser radar systems, can be used for large area initial search. After locating the target within the angular resolution of these systems, the angular position resolution can be improved by pointing the laser radar in the direction of the target and tracking the target. Additionally, since microwave and millimeter-wave systems operate in different frequency regions than the laser radars, this dual sensor fusion can be used to eliminate false targets due to system noise or external clutter.

In this chapter the general theory of operation of MMW radars is given. Particular emphasis has been placed in the utilization of these radars to detect nonmoving targets and flight path obstacles. Two MMW radars (35 and 95 GHz) are used for numerical examples to illustrate the application of the covered principles.

4.1 DETECTION OF GROUND TARGETS

The radar search range equation for a matched transmit-receive system (from Chapter 2) can be written as follows:

$$R = \left[\frac{\overline{P} A \sigma}{(16)(kT)(L)(\text{S/N})} \frac{t_s}{\Omega} \right]^{1/4} \tag{4.1}$$

or in terms of S/N we have

$$(\text{S/N})_t = \frac{\overline{P} A \sigma}{(16)(kT)(L)R^4} \left(\frac{t_s}{\Omega} \right) \tag{4.2}$$

where

$\overline{P}$ = average power
= $P\tau$ PRF
P = peak power
τ = pulse duration
PRF = pulse repetition frequency
A = antenna aperture
σ = radar target cross section
k = Boltzmann constant
T = equivalent noise temperature
L = system losses
R = radar-target range
t_s = radar scan time
Ω = radar angular search coverage

In stationary ground target applications, when the antenna mainbeam is focused on the ground, as shown in Figure 4.1, the ground resolution or the clutter return cell will be

$$\text{resolution cell} = (R\theta_{\text{BW}})(c\tau/2) \tag{4.3}$$

where θ_{BW} is the antenna beamwidth. Note that the $c\tau/2$ of this equation is a function of the radar depression angle ϵ. In radar-bearing helicopter applications this angle is very small, making Eq. (4.3) correct as written. The actual clutter radar cross section, however, will be the resolution cell of Eq. (4.3) multiplied by backscattering cross section of the clutter area. This backscattering cross section, usually designated by σ_0 is a measured value and can be obtained from [5] for MMW radars. Using this in Eq. (4.3) we get the clutter radar cross section:

$$\sigma_c = (R\theta_{BW})(c\tau/2)\sigma_0 \tag{4.4}$$

Substituting Eq. (4.4) in S/N ratio Eq. (4.2), we get

$$(\mathrm{S/N})_c = \frac{\overline{P}A\theta_{BW}c\tau\sigma_0}{32(kT)LR^3}\left(\frac{t_s}{\Omega}\right) \tag{4.5}$$

Equation (4.5) represents the clutter-signal-to-noise ratio against which the target signal of a ground target, such as a tank or a truck, should compete for detection. Note that the ground clutter-signal-to-noise ratio is inversely proportional to R^3 rather than R^4 of target signal power as given in Eq. (4.2).

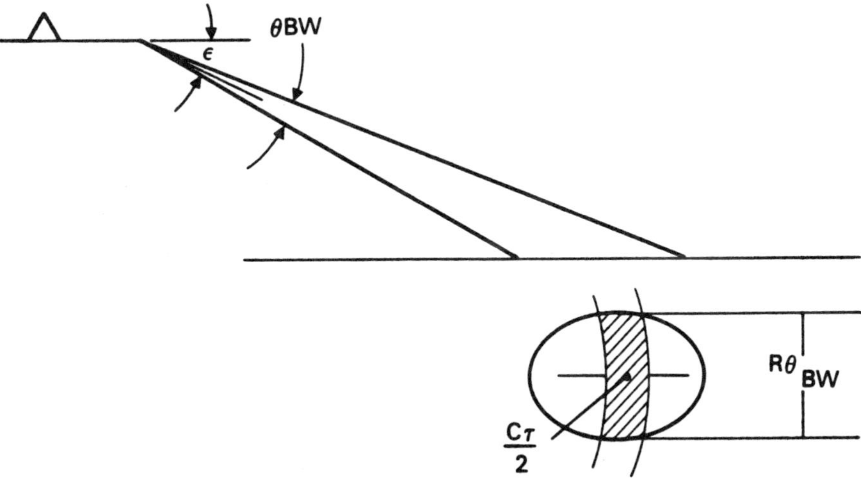

Figure 4.1 Airborne radar with ground resolution cell.

Rain backscattering *radar cross section* (RCS) can be computed by calculating the volume occupied by rain and multiplying this volume by appropriate backscattering coefficient. The maximum rain backscattering volume will depend on the antenna beam area coverage and range gate length; that is:

$$\sigma_r = \frac{\pi}{4}(R\theta_{BW})^2\left(\frac{c\tau}{2}\right)\sigma_i \tag{4.6}$$

where σ_i is the rain backscattering coefficient, a measured number. This value can be obtained from [6].

Using the value of σ_r from Eq. (4.6) in Eq. (4.2) for the radar cross section, σ, we get

$$(\mathrm{S/N})_r = \frac{\overline{P}A\pi\theta_{BW}^2 c\tau\sigma_i}{128(kT)(L)R^2}\left(\frac{t_s}{\Omega}\right) \tag{4.7}$$

Equation (4.7) represents the rain S/N ratio against which the target signal should compete for detection. The target signal of a stationary ground target should exceed the ground clutter signal for detection against ground clutter. It should also exceed the rain backscattering signal to be detected under adverse weather conditions. Note that the S/N ratio from rain backscattering is inversely proportional to R^2 rather than R^4 of target signal power.

4.1.1 Atmospheric Attenuation

In addition to the rain backscattering and ground clutter effects, atmospheric attenuation, which accounts for the loss of electromagnetic energy in traveling through the atmosphere, should also be included in target detection and S/N ratio calculations. Atmospheric attenuation, and for that matter also rain backscattering, increases with increasing frequency. That is, both atmospheric attenuation and backscattering coefficients are generally less at 35 GHz than at 95 GHz. Table 4.1 gives these values obtained from Reference [6].

Atmospheric attenuation values as seen from Table 4.1 are given in units of dB/km and increase linearly (in dB) with range. Since dBs are logarithmic units, the atmospheric loss as a function of range will be exponential. Inclusion of this loss in the radar range equations, (4.2), (4.5), and (4.7), will result in transcendental equations that are difficult

Table 4.1 Atmospheric Attenuation and Rain Backscattering at 35 and 95 GHz

Frequency	*35 GHz*	*95 GHz*
Clear air		
attenuation	0.18 dB/km	0.24 dB/km
Fog (thick 0.1 g/m^3)		
attenuation	0.06 dB/km	0.35 dB/km
Light rain (1 mm/h)		
attenuation	0.24 dB/km	0.95 dB/km
backscattering	0.21 cm^2/m^3	0.89 cm^2/m^3
Moderate rain (4 mm/h)		
attenuation	1 dB/km	2.8 dB/km
backscattering	1.2 cm^2/m^3	2 cm^2/m^3
Heavy rain (16 mm/h)		
attenuation	4 dB/km	7.4 dB/km
backscattering	4.9 cm^2/m^3	3.9 cm^2/m^3

to solve. There are, however, other methods of including atmospheric attenuation effects in the detection range computation, and these will be described.

The graphical method of including atmospheric attenuation consists initially of S/N ratio computation (without atmospheric attenuation) as a function of range using Eqs. (4.2), (4.5), and (4.7), and plotting a graph of S/N (in dB) *versus* range. Then, using this graph, at each value of range R, the two-way atmospheric attenuation in dB is subtracted from the corresponding S/N value. This new curve of S/N *versus* range (which includes atmospheric attenuation) becomes the basis of detection range computation. This method is described through a numerical example in Chapter 6 for the case of a laser radar system.

Another method of including atmospheric attenuation effects, which may be more convenient for digital computer solution, involves tabulating detection ranges with atmospheric attenuation and subsequently computing detection ranges without atmospheric attenuation. Having this table, and the computed value of detection range *without* atmospheric attenuation from Eq. (4.2), we can find the corresponding detection range that includes atmospheric attenuation. This method, which is also used in Chapters 5 and 6 for the electro-optical calculations, will be described in the second part of this chapter through a numerical example.

4.1.2 Computational Procedure

For specified radar and target parameters, digital computer programs can be written to compute target, clutter, and rain backscattering S/N as a function of range as given by Eqs. (4.2), (4.5), and (4.7). The methods of including atmospheric attenuation in performance calculations have been discussed and will be illustrated in Chapter 6. The results of one such calculation are given in Figure 4.2. From this figure the target detection range can be obtained for any given S/N ratio. For example, a target-to-clutter-signal power requirement of 10 dB will result, from Figure 4.2, at a detection range of about 1.4 km. By specifying a detection S/N ratio, the procedure depicted in Figure 4.2 can be programmed for digital computer solution.

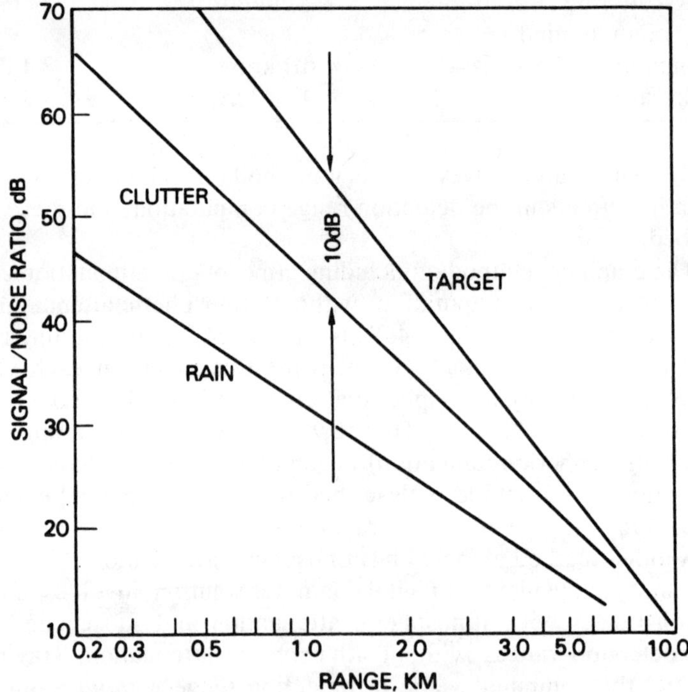

Figure 4.2 Target, clutter, and rain backscattering S/N ratio as a function of range.

4.1.3 Numerical Examples: Stationary Ground Targets

The detection range calculations just discussed were made for two MMW radars, 35 and 95 GHz, with the parameters given in Table 4.2. These radar parameters are used in conjunction with atmospheric attenuation and rain backscattering values given in Table 4.1. Most of the parameters of Table 4.2 are self-explanatory; the following, however, is the reason for some of these selections.

Table 4.2 MMW Radar Parameters of Example Problem

Item	*Symbol*	*35 GHz*	*95 GHz*
Peak power, W	P	20	20
Pulse width, s	τ	50E − 9	50E − 9
PRF, Hz	PRF	10000	10000
RCS, m^2	σ	30	30
Antenna aperture, m^2 (12-in diameter)	A	0.073	0.073
Radar frametime, s	t_s	3.6	4.1
Equivalent noise temp, K	T	1170	1470
Losses, dB	L	9	13
Azimuth scan	Az	60°	60°
Elevation scan	El	4	4
Antenna beamwidth	θ_{BW}	1.60°	0.60°
Ground backscattering coefficient	σ_0	0.01	0.01

The pulse width of 50E − 9 s will result in a range resolution cell ΔR of

$$\Delta R = \frac{c\tau}{2} = \frac{3 \times 10^3 \times 50E - 9}{2} = 7.5 \text{ m} \quad (4.8)$$

The average power of the radar can be computed from

$$\bar{P} = P\tau \, \text{PRF} = 20 \times 50E - 9 \times 10000 = 0.01 \text{ W} \quad (4.9)$$

In this example, the radar target cross section of the ground target is taken as 30 m^2. The antenna for both 35 and 95 GHz systems is considered to be a circular dish with a diameter of 12 in. The antenna beamwidth is calculated from λ/D where λ is the wavelength and D is the diameter and given in the table. The radar frame time, which is the time to cover the azimuth and elevation volume coverage of $60 \times 4°$, is given as 3.6 s for 35 GHz radar and 4.1 s for the 95 GHz radar. With TV raster scan, the 95 GHz radar's smaller beamwidth requires more elevation bars, thereby a greater amount of turn-around time. The equivalent noise temperature is computed from receiver noise temperature of 293 K by using a 6 dB noise figure for the 35 GHz radar and a 7 dB figure for the 95 GHz radar. System losses of 35 GHz are considered to be 9 dB and that of 95 GHz are considered to be 13 dB. The increased loss of 95 GHz radar is primarily due to the higher transmit-receive losses occurring at 95 GHz frequency as compared to 35 GHz. Note that atmospheric attenuation losses are not included in these system loss values. Atmospheric attenuation losses are calculated separately, as discussed earlier.

The ground backscattering coefficient of 0.01 is obtained from Reference [5]. This value is for small depression angles for deciduous trees and is averaged over wet and dry conditions.

4.1.4 Numerical Results

The parameters of Table 4.2 were used in appropriate computer programs with the detection ranges obtained and tabulated in Table 4.3. Note that from this table the 95 GHz radar results in greater detection ranges against stationary ground targets than the 35 GHz radar. This is primarily due to the smaller antenna beam and the resulting reduced clutter signal. In clear weather the range multiplier of 95 and 35 GHz is approximately 3, and this is the same as the antenna beam multiplier. Under rainy conditions, however, the multiplier becomes considerably smaller primarily because of higher atmospheric attenuation of 95 GHz radar as compared to the 35 GHz.

For the purpose of comparison Table 4.4 gives ranges of 95 and 35 GHz MMW radars for detection of moving targets; that is, using *moving target indication* (MTI). In this case the target return energy competes only with system noise, rather than clutter or rain backscattering. The values of Table 4.4 are computed using the MMW radar parameters in search range equation (4.1) and incorporating the atmospheric attenuation

Table 4.3 Ranges (in km) of 95 and 35 GHz Radar in Detection of Stationary Ground Targets

Condition	*35 GHz*	*95 GHz*
Clear	1.5	4
Fog	1.7	4
Light Rain	1.5	4
Moderate Rain	1.2	3
Heavy Rain	1.1	1.5

Table 4.4 Ranges (in km) of 95 and 35 GHz Radars in Detection of Moving Targets

Condition	*35 GHz*	*95 GHz*
Clear	12	9
Fog	14	8
Light Rain	11	6
Moderate Rain	7	3.5
Heavy Rain	3.5	2

losses of Table 4.1. Using the parameters of the 35 GHz radar example in Eq. (4.1) we get

$$R = \left[\frac{(0.01)(0.073)(30)(3.6)(57.3)^2}{(16)(1.38 \times 10^{-23})(1170)(7.94)(60 \times 4)(10)}\right]^{1/4} = 15 \text{ km} \qquad (4.10)$$

where an S/N of 10 is assumed for detection. Similarly, using the parameters of 95 GHz radar, we get

$$R = \left[\frac{(0.01)(0.073)(30)(4.1)(57.3)^2}{(16)(1.38 \times 10^{-23})(1470)(20)(60 \times 4)(10)}\right]^{1/4} = 12 \text{ km} \qquad (4.11)$$

Incorporating the atmospheric attenuation values of Table 4.1 in detection ranges of 35 and 95 GHz radar will result in the detection values of Table 4.4. Note that although power-aperture product of the two radars are the same, the 35 GHz radar, because of lower atmospheric attenuation, results in better moving target detection ranges. Also note that to obtain range R for moving targets, where the primary measurement is closing rate, $\dot{R}$ (doppler or frequency shift), will involve special waveform design, such as frequency modulation ranging (Chapter 2). The transmission and

and reception of this waveform will occupy additional on-target time, effectively reducing the detection ranges given in Table 4.4.

Figure 4.3 summarizes the results of this study by giving the values of detection ranges of stationary and moving targets for a variety of weather conditions. It is seen that moving targets can be detected at longer ranges by 35 GHz rather than by 95 GHz radars. Stationary targets, however, are detected by 95 GHz radars at longer ranges. Note that, as discussed earlier, the ranges for moving targets of Figure 4.3 do not include waveform design losses. These losses will incur in obtaining radar-target *range* in addition to the moving target indication. It is estimated that the incorporation of ranging waveform in the radar may reduce moving target detection ranges by as much as 20 percent.

It is interesting to note that the results of this chapter are in general agreement with the corresponding results by V. L. Lynn given in Reference [6], Table 2, page 625. The radar performance results of the reference were arrived at by completely different considerations. The maximum range values of this table give a ratio of 2 : 1 between 35 and 95 GHz radars. The fixed target maximum detection ranges of this table give a ratio of 3 : 1 between 95 and 35 GHz radars. These ratios most closely approximate those of our Tables 4.4 and 4.3, respectively, for the moderate rain condition.

Furthermore, it should be observed that the angular position accuracy of the 95 GHz radar in the search mode of operation will be approximately one-half the beamwidth of 0.60° or 0.30°. The corresponding value for 35 GHz radar will be 0.80°. Accuracy of these angular measurements obtained in the search mode of operation can be improved considerably by continuous tracking of the target in the tracking mode of operation. Target tracking accuracies are given in Chapter 2.

4.2 DETECTION OF HOVERING HELICOPTERS AGAINST A HILLY BACKGROUND

Consider the detection of hovering helicopters against a hilly background by an airborne radar, the geometry of which is shown in Figure 4.4. The mainbeam clutter area in this case can be approximated by

$$A_c = \frac{\pi D_a^2}{4} \tag{4.12}$$

where A_c is the clutter area and D_a is the diameter of the mainbeam coverage of the area. The value of D_a can be approximated by using

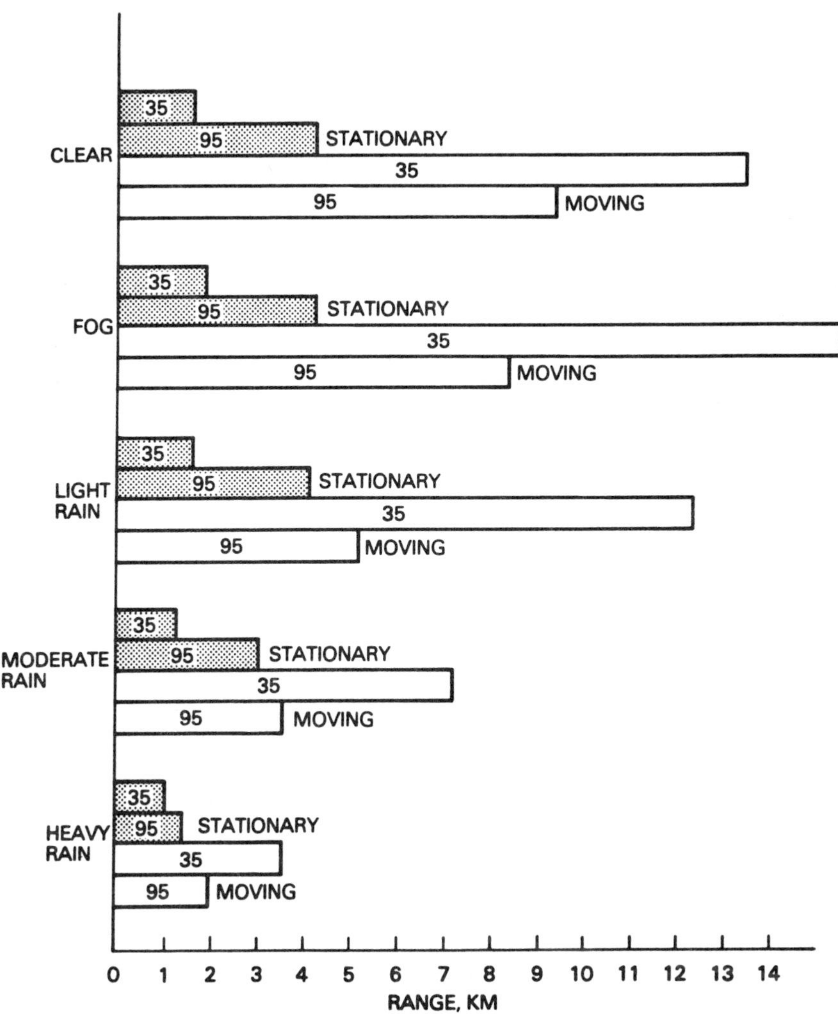

Figure 4.3 Detection ranges of stationary and moving ground targets for 35 and 95 GHz MMW radars for a 1-ft antenna dish.

radar-clutter range R and antenna half-power beamwidth θ_{BW}; that is:

$$D_a = R\theta_{BW} \tag{4.13}$$

Using this value in Eq. (4.12) and designating the clutter backscattering coefficient by σ_0, we get the equivalent clutter radar cross section as

$$\sigma_c = \frac{\pi}{4}(R\theta_{\mathrm{BW}})^2\sigma_0 \tag{4.14}$$

Using this in the S/N ratio equation, (4.2) for the radar target cross section, σ, we get

$$(\mathrm{S/N})_c = \frac{\pi \overline{P} A \theta_{\mathrm{BW}}^2 \sigma_0}{64(kT)LR^2}\left(\frac{T_s}{\Omega}\right) \tag{4.15}$$

Note that targets cannot be detected against mainlobe clutter, because of the high backscattering value of the terrain perpendicular to the mainbeam of the antenna. Targets can, however, be detected several range gates away from the intersection of the mainbeam and the terrain, in the sidelobe clutter region. Allowing a two-way sidelobe clutter reduction factor of S_c, the S/N equation, (4.15), can be written

$$(\mathrm{S/N})_c = \frac{\pi \overline{P} A \theta_{\mathrm{BW}}^2 \sigma_0}{64(kT)LR^2 S_c}\left(\frac{t_s}{\Omega}\right) \tag{4.16}$$

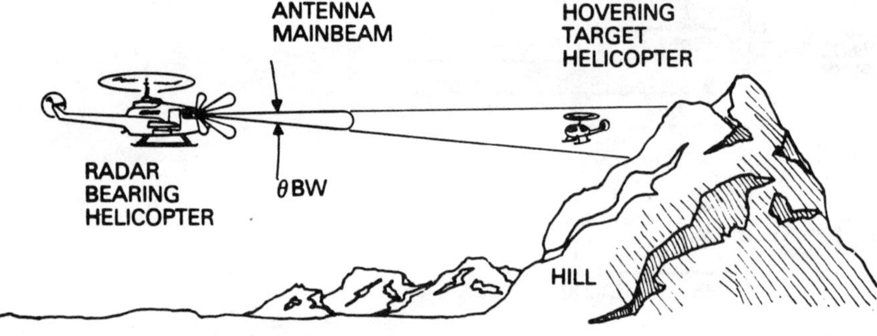

Figure 4.4 Geometry of a hovering helicopter against a hilly background.

This equation, derived for a hovering helicopter against hilly background, is equivalent to Eq. (4.5), derived for a stationary ground target. The S/N ratio of Eq. (4.16) is proportional to the square of radar target range R. The sidelobe reduction parameter absent in Eq. (4.5) is introduced in (4.16). Also note that the ground backscattering coefficient, σ_0, of Eq. (4.16) will be considerably greater than that of Eq. (4.5) primarily because of the angle of incidence. In the derivation of Eq. (4.16) it is assumed that the target helicopter is only a few range gates away from the mainlobe clutter region. For this reason, no distinction is made between the radar-target and radar-clutter range R.

4.2.1 Numerical Results

The detection ranges of hovering helicopters against a hilly background were calculated using the radar parameters of Table 4.2, with the following additional numerical values:

$$\begin{aligned} &\text{ground backscattering } \sigma_0 = 1.0\,(0\text{ dB}) \\ &\text{two-way sidelobe–mainlobe ratio, } S_c = 1000\,(30\text{ dB}) \end{aligned} \tag{4.17}$$

It is further assumed that the entire mainbeam is intercepted by the hilly background. The backscattering coefficient of Eq. (4.17) is for a perpendicular angle of incidence. The sidelobe reduction of Eq. (4.17) is equivalent to a one-way sidelobe of 15 dB below the mainlobe. The radar cross section of the target helicopter is assumed to be 30 m^2. Using these discussed parameters, together with the outlined calculation procedure, we obtain the detector ranges for hovering helicopters given in Table 4.5.

Table 4.5 Ranges (in km) of 95 and 35 GHz Radar in Detection of a Hovering Helicopter against a Hilly Background

Condition	*35 GHz*	*95 GHz*
Clear	2.2	6
Fog	2.2	6
Light rain	2.2	6
Moderate rain	2.2	3.6
Heavy rain	1.2	2.1

The detection values of Table 4.5 are similar to those of Table 4.3, for detection of stationary ground targets. The 3 : 1 frequency ratio between 35 and 95 GHz radars applies also to the given detection ranges for clear weather conditions. This ratio, however, decreases in adverse weather, as rain attenuation values under adverse weather conditions are considerably greater for 95 GHz than for 35 GHz radars. Note that since the radar parameters remain essentially the same as for detection of moving ground targets, the moving (rather than hovering) helicopter detection ranges will be the same as given in Table 4.4. The stationary and moving target detection ranges are shown in Figure 4.5.

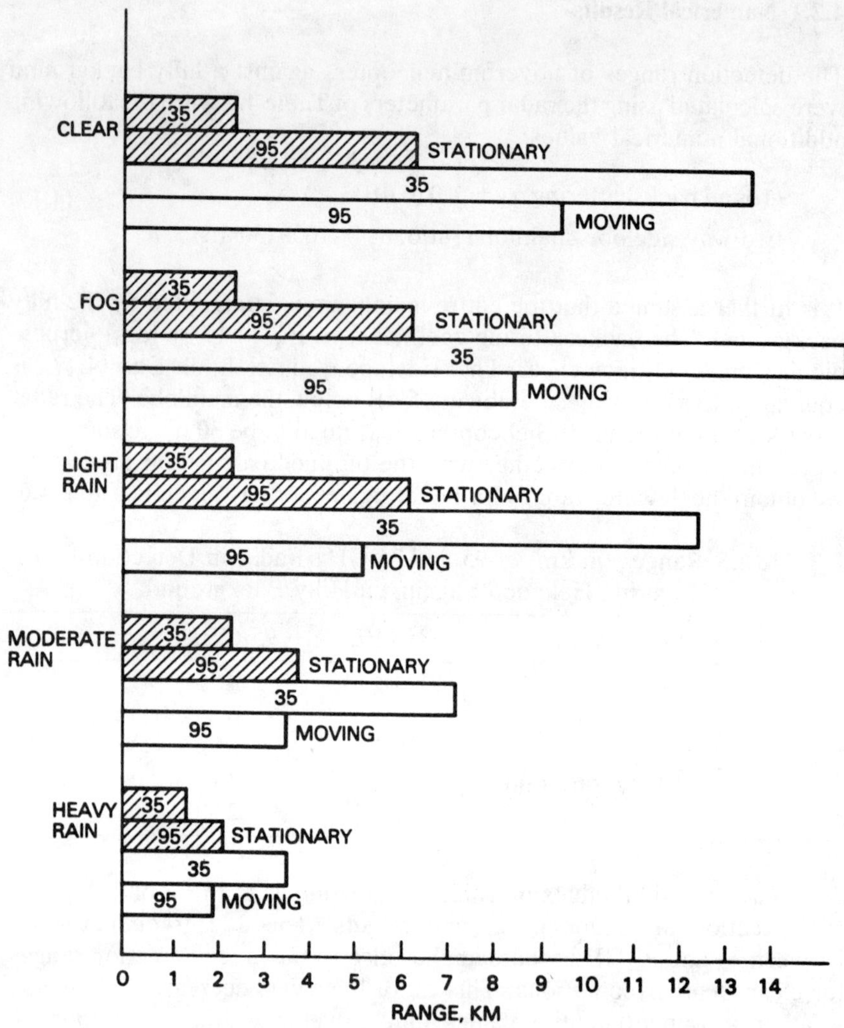

Figure 4.5 Detection ranges of hovering and moving helicopters for 35 and 95 GHz MMW radars for a 1-ft antenna dish.

4.3 DISCUSSION OF NONMOVING TARGET DETECTION

The difference between the detection ranges of 35 GHz and 95 GHz radars can be analytically derived from a consideration of S/N ratio Eqs. (4.2), (4.5), and (4.16). In the case of stationary ground targets, dividing the S/N ratio equation (4.2) by the clutter-to-noise ratio equation (4.5), we can find the target-signal-to-clutter signal ratio as

$$\frac{S_t}{S_c} = \frac{2\sigma}{(\tau c)\theta_{BW} R_t \sigma_0} \tag{4.18}$$

Note that this equation does not include atmospheric effects. For a required target-to-clutter signal ratio for detection, Eq. (4.18) can be solved for target range, R_t, as

$$R_t = \frac{2\sigma}{(\tau c)\theta_{BW}(S_t/S_c)\sigma_0} \tag{4.19}$$

From Eq. (4.19) we see that the target detection range is inversely proportional to range gate size $(\tau c/2)$, antenna beamwidth θ_{BW}, and backscattering coefficient σ_0. Since for *a given antenna size* the 95 GHz radar will have a smaller beamwidth, its detection range will be accordingly higher.

In a similar fashion, dividing Eq. (4.2) by (4.7) will result in target-signal-to-rain-backscattering signal as

$$\left(\frac{S_t}{S_B}\right) = \frac{8\sigma}{(\tau c)\theta_{BW}^2 \pi \sigma_i R_t^2} \tag{4.20}$$

For a given (S_t/S_B) detection ratio Eq. (4.20) can be put in terms of R_t as follows:

$$R_t^2 = \frac{8\sigma}{(\tau c)\theta_{BW}^2 \pi \sigma_i (S_t/S_B)} \tag{4.21}$$

From Eq. (4.21) we see that the square of detection range of a target in rain is inversely proportional to the range gate value $(\tau c/2)$, the antenna beam area $(\simeq \pi\theta_{BW}^2)$, and the rain backscattering coefficient σ_i.

In the case of hovering helicopters, dividing Eqs. (4.2) and (4.16) will result in a target-to-clutter signal ratio of

$$\frac{S_t}{S_c} = \frac{4\sigma S_1}{\pi\theta_{BW}^2\sigma_0 R^2} \tag{4.22}$$

For a required target-signal-to-clutter-signal ratio, Eq. (4.22) can be solved for range R:

$$R^2 = \frac{4\sigma S_1}{\pi\theta_{BW}^2\sigma_0(S_t/S_c)} \tag{4.23}$$

From Eq. (4.23) we see that detection range R is inversely proportioned to antenna beamwidth θ_{BW}, or it is directly proportional to transmit frequency. As seen in Table 4.5, higher frequencies generally achieve longer detection ranges.

We can see that the detection range trends in clutter and rain depicted by Eqs. (4.19), (4.21), and (4.23) apply in cases where the target, clutter, or rain signals are above the system noise level. For this reason, in calculating actual detection ranges, the complete detection range equations given by Eqs. (4.2), (4.5), (4.7), and (4.16) should be used.

4.4 DETECTION OF WIRES AND POLES BY MMW RADARS

A recent article in *Aviation Week and Space Technology* [7] described a 95 GHz MMW radar system developed by Thomson CSF Company for detecting obstacles in helicopter flight. This radar operating in the search mode of operation can produce a display similar to that of Figure 4.6. This figure shows a display of tension wires that may constitute a hazard to the helicopter flight.

In this section of the report we demonstrate that both 35 and 95 GHz MMW radars can be useful in detecting wires and wooden and metallic poles in the flight path of helicopters. Minimum detection ranges of 1.1 km were achieved for the specified radar parameters for clear weather and fog. Because of rain backscattering effects, however, off-axis wire and wooden pole detections became nearly impossible in rain.

4.4.1 Radar Cross Section of Wires

Assuming wires consist of conducting material, their radar cross section σ perpendicular to the axis of the wire can be computed from [4] (page 52), as follows:

$$\sigma = 2\pi a L^2/\lambda \tag{4.24}$$

where a is the radius of the wire, λ the wavelength, and L the length of the wire. Using Eq. (4.24), a wire $\frac{1}{2}$ in in diameter and 10 m long will have the following radar cross sections:

$$\begin{aligned} \sigma\,(35\ \text{GHz}) &= 4.652\text{E}2\ \text{m}^2 \\ \sigma\,(95\ \text{GHz}) &= 1.266\text{E}3\ \text{m}^2 \end{aligned} \tag{4.25}$$

Figure 4.6 Display of high tension wires from MMW radar data in helicopter flights.

The radar cross section of edges, metal strips and long thin wires for off-normal angles is given in [8] as follows:

$$\sigma = \frac{\pi (L \sin\theta)^2 \left(\dfrac{\sin(2\pi L \cos\theta/\lambda)}{2\pi L \cos\theta/\lambda} \right)}{(\pi/2)^2 + \ln^2 \left(\dfrac{\lambda}{\gamma \pi a \sin\theta} \right)} \tag{4.26}$$

where

σ = radar cross section
L = length of the wire
θ = angle between the incident wave and the axis of the wire
γ = Eulers constant (1.78)
a = radius of the wire

Equation (4.26) was programmed for a computer solution and the values of σ were obtained and plotted as a function of off-normal angle, as given in Figure 4.7. From this figure we see that RCS values are at a maximum perpendicular to the wire (zero off-normal angle) and reduce from this value with an increasing off-normal angle.

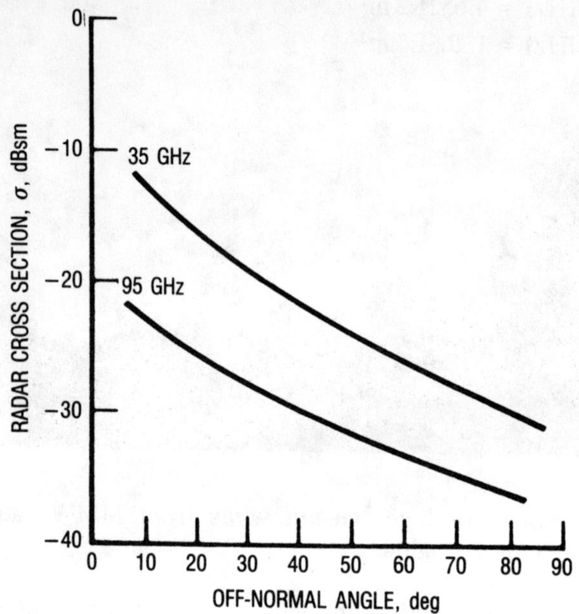

Figure 4.7 Radar cross section of $\frac{1}{2}$-in wire as a function of off-normal angle for 35 and 95 GHz radars.

4.4.2 Radar Cross Section of Poles

The RCS of wooden poles can be computed using area of the poles and the backscattering coefficient of trees from measured data. Assuming that at a given time a 5 m length of the pole is exposed to the radar, and the pole has a width of 1 ft, the actual cross section of the pole will be 1.52 m^2. From [6] (page 228) the backscattering cross section of trees at 95 GHz is given as -28 dB (1.58E $-$ 3). Using this number the RCS of the wooden pole becomes

$$\begin{aligned} \text{RCS} &= (1.52)(1.58\text{E} - 3) \\ &= 2.41\text{E} - 3 \text{ m}^2 \end{aligned} \tag{4.27}$$

Considering that poles contain metallic parts and connectors, as opposed to trees, the overall backscattering coefficient will be higher than −28 dB and can be estimated at −20 dB. With this value the RCS of wooden poles becomes

$$\text{RCS} = (1.52)(.01) = 0.0152\ \text{m}^2 \tag{4.28}$$

Because of the absence of backscattering data at 35 GHz frequency, the radar cross section of wooden poles at this frequency is also assumed to be equal to the value given in Eq. (4.28).

The radar cross section of metallic poles can be computed using RCS equations of a cylinder. This equation is exactly the same as that of a wire given in Eq. (4.24). Using a diameter of 9 in and a length of 5 m, Eq. (4.24) for 35 GHz and 95 GHz radars results:

$$\begin{aligned} \text{RCS (35 GHz)} &= 2.09\text{E}3\ \text{m}^2 \\ \text{RCS (95 GHz)} &= 5.70\text{E}3\ \text{m}^2 \end{aligned} \tag{4.29}$$

These values are considerable and will make the detection of metallic poles relatively simple.

4.5 RADAR DETECTION RANGES OF OBSTACLES

The detection range of a matched radar system in the search mode of operation was given in Eq. (4.1):

$$R^4 = \frac{\overline{P}A\sigma}{(16)\,kTL(\text{S/N})}\left(\frac{t_s}{\Omega}\right) \tag{4.30}$$

For numerical calculations consider the radar parameters of 35 and 95 GHz radars given in Table 4.2. In this application, however, let us increase the peak power to 60 W and assume a required detection signal to noise ratio of 10 dB. The average power in this case becomes

$$\begin{aligned} \overline{P} &= P\tau\ \text{PRF} \\ &= 0.03\ \text{W} \end{aligned} \tag{4.31}$$

Using the radar parameters of Table 4.2 with these modifications, we get for 35 GHz radar,

$$R^4 = 4.835 \times 10^{15}\sigma \quad \text{(range in m)} \tag{4.32}$$

And, similarly, for 95 GHz radar we will have

$$R^4 = 1.909 \times 10^{15}\sigma \quad \text{(range in m)} \tag{4.33}$$

where σ is the radar target cross section.

4.5.1 Atmospheric Attenuation and Rain Backscattering

Rain backscattering and atmospheric attenuation, which accounts for the loss of electromagnetic energy in traveling through the atmosphere, should be included in target detection and S/N ratio calculations. Atmospheric attenuation, and for that matter also rain backscattering, increase with increasing frequency. That is, both atmospheric attenuation and backscattering coefficients are less at 35 GHz than at 95 GHz. Table 4.1 lists atmospheric attenuation and rain backscattering values for 35 and 95 GHz radars.

As described previously, atmospheric attenuation loss can be included in detection range calculation either graphically or by a tabular method that matches the calculated "no loss" detection range to the corresponding detection range, which includes atmospheric attenuation loss. As a numerical example of this procedure, consider that a "no loss" detection range of 39 km has been computed for a 35 GHz MMW radar. We are required to compute the detection range of this radar under moderate rain conditions. From Table 4.1, the atmospheric attenuation for this condition is 1 dB/km.

Having this value, we compute the numerical values of Table 4.6. The first column is range *with* atmospheric loss in km. The second column is the corresponding two-way atmospheric loss in dB. The third column is the range equivalent of this loss. Since S/N and range have a fourth power relationship, the equivalent range factor—for example, for 4 dB (2.511)—will be the fourth root of this number, or 1.26 as given in the second row of the table. Multiplying the range factor by the original (with loss) range values, we get the equivalent "no loss" range as given in the fourth column of the table. Thus, the "no loss" range of 39 km given in the problem, will correspond to an actual detection range of 11 km under moderate rain condition, as given in the table.

The rain backscattering will increase the noise level of the radar system. This effect can be included in the calculation through computing the ratio of target signal power to rain backscattering signal power. This will be shown through numerical examples.

Table 4.6 Computation of Range with Atmospheric Attenuation Given the "No Loss" Detection Range

Range with Loss (km)	*Attenuation Loss (dB)*	*Range Factor*	*Range without Loss (km)*
1	2	1.12	1.12
2	4	1.26	2.52
3	6	1.41	3.42
4	8	1.58	6.32
5	10	1.78	8.90
6	12	2.00	12.00
7	14	2.24	15.68
8	16	2.51	20.08
9	18	2.81	25.29
10	20	3.16	31.60
11	22	3.54	39.02
12	24	3.98	47.77

4.5.2 Detection Range Calculations

Using the RCS values of perpendicular to wire case from Eq. (4.25) for 35 and 95 GHz radars, together with Eqs. (4.32) and (4.33) we get

$$\begin{aligned} R\,(35\text{ GHz radar}) &= 38.7\text{ km} \\ R\,(95\text{ GHz radar}) &= 39.44\text{ km} \end{aligned} \tag{4.34}$$

The detection ranges of Eq. (4.34) should be calculated including the atmospheric attenuation losses of Table 4.1. Using these values will obtain the detection ranges of Table 4.7. The method of including atmospheric attenuation in detection range calculation was discussed previously and illustrated through numerical example of Table 4.6. In fact, this numerical example corresponds to the moderate rain case of Table 4.7.

Note that the values of Table 4.7 include atmospheric attenuation only. In addition to attenuation, rain backscattering should also be calculated. The ratio of target power to rain backscattering power can be represented by Eq. (4.20) as follows:

$$\left(\frac{S_t}{S_r}\right) = \frac{8\sigma}{(\tau c)\,\theta_{\text{BW}}^2 \pi \sigma_i R_t^2} \tag{4.35}$$

where θ_{BW} is the antenna beamwidth and σ_i is the backscattering coefficient from Table 4.1, and c is the velocity of propagation. For example, under a heavy rain condition Eq. (4.35) for a 95 GHz will result in

$$\left(\frac{S_t}{S_r}\right) = \frac{8(1.266\text{E}3)}{(50\text{E}-9)(3\text{E}8)(0.60/57.3)^2\pi(3.9\text{E}-4)(3000)^2}$$
$$= 571$$
$$= 27.5\text{dB} \quad (4.36)$$

This means that the target return power in the particular range gate is higher by 27.5 dB than the rain backscattering power. In a similar manner Eq. (4.35) can be applied to other detection ranges both for 35 and 95 GHz radars. These results will show that target power always will exceed rain backscattering power, and detection ranges of Table 4.7 can be considered to be the realizable detection ranges that include rain backscattering.

Table 4.7 Perpendicular to Wire Detection Ranges (in km) for 35 and 95 GHz MMW Radars

Condition	*35 GHz*	*95 GHz*
Clear Air	23	22
Fog	31	19
Light rain	21	10
Moderate rain	11	6
Heavy rain	4.5	3

4.5.3 Nonperpendicular to Wire Detection Ranges

From a consideration of the RCS values of Figure 4.7, let us assume the following RCS values for 30–50° off-normal angles:

$$\text{RCS (35 GHz radar)} = 8 \times 10^{-3}\,\text{m}^2$$
$$\text{RCS (95 GHz radar)} = 1.5 \times 10^{-3}\,\text{m}^2 \quad (4.37)$$

Using these in the range equations (4.32) and (4.33), we get

$$R\,\text{(35 GHz radar)} = 2.5\,\text{km}$$
$$R\,\text{(95 GHz radar)} = 1.3\,\text{km} \quad (4.38)$$

Using these nonattenuated detection ranges, Table 4.8 gives MMW detection ranges that include atmospheric attenuation. Because of the short distances involved and the low attenuation values in the clear and foggy weather, the attenuated detection ranges are not that different from the nonattenuated ones. The rain backscattering check of Eq. (4.35) for light, moderate, and heavy rain using the detection ranges of 95 GHz radar from Table 4.8, together with atmospheric propagation values, will result in

$$\begin{aligned} \left(\frac{S_t}{S_r}\right)_{\text{light}} &= -16.47\,\text{dB} \\ \left(\frac{S_t}{S_r}\right)_{\text{moderate}} &= -29\,\text{dB} \\ \left(\frac{S_t}{S_r}\right)_{\text{heavy}} &= -20.8\,\text{dB} \end{aligned} \tag{4.39}$$

From these values, we see that rain backscattering will make the off-normal wire detection ranges of 95 GHz MMW radar practically non-existent during rainy weather.

Similarly, we can see, by using Eq. (4.35), that rain backscattering will also interfere with off-normal wire detection using 35 GHz radar. Thus, both 35 and 95 GHz MMW radars will be unable to detect wires under the off-normal look conditions in rainy weather.

Table 4.8 Off-Normal Wire Detection Ranges (in km) for 35 and 95 GHz MMW Radars

Condition	*35 GHz*	*95 GHz*
Clear Air	2.4	1.1
Fog	2.5	1.1
Light rain	2.4	1.0
Moderate rain	2.0	0.9
Heavy rain	1.3	0.7

4.5.4 Wooden Pole Detection Ranges

The detection range of wooden poles by MMW radars can be computed using the RCS values of Eq. (4.28) in the radar range equations (4.32)

and (4.33) as follows:

$$R\,(35\text{ GHz radar}) = 2.93\text{ km} \qquad (4.40)$$
$$R\,(95\text{ GHz radar}) = 2.32\text{ km}$$

Using the atmospheric attenuation values of Table 4.1 the detection ranges of Eq. (4.40) will decrease to the values shown in Table 4.9.

Table 4.9 Detection Ranges (in km) of Wooden Poles by 35 and 95 GHz MMW Radars

Condition	*35 GHz*	*95 GHz*
Clear Air	2.8	2.2
Fog	3.0	2.2
Light rain	2.8	1.9
Moderate rain	2.3	1.5
Heavy rain	1.5	1.0

The rain backscattering ratio check of Eq. (4.35) will show that the target return power during rain is considerably smaller than rain backscattering power. This will effectively eliminate the detection capability of MMW radar in the detection of wooden poles in rainy weather.

4.5.5 Metallic Pole Detection Ranges

The MMW radar detection range of metallic poles can be determined by using the RCS values of Eq. (4.29), together with the radar range equations, (4.32) and (4.33), as follows:

$$R\,(35\text{ GHz radar}) = 56.38\text{ km} \qquad (4.41)$$
$$R\,(95\text{ GHz radar}) = 57.43\text{ km}$$

Using these values with the atmospheric attenuation factors of Table 4.1, we obtain the detection ranges given in Table 4.10.

We can see that the rain backscattering effect on the detection values is minimal, and these values can be considered to be actual detection ranges of the MMW radars against metallic poles.

Table 4.10 Detection Ranges (in km) of Metallic Poles by 35 and 95 GHz MMW Radars

Condition	*35 GHz*	*95 GHz*
Clear Air	30	27
Fog	42	23
Light rain	27	13
Moderate rain	13	7
Heavy rain	5	3

4.6 SUMMARY

We have seen that 35 and 95 GHz radars in the search mode of operation can be useful tools for detecting nonmoving targets, as well as obstacles such as wires and poles, under adverse weather conditions or at night. Millimeter-wave radars operating at 95 GHz frequency can detect nonmoving targets on the ground, such as tanks and trucks, and in the air, such as hovering helicopters, at longer ranges than 35 GHz radars by a factor of 3 : 1. This ratio is almost the same as the corresponding frequency ratio. The detection range of moving targets, however, is longer for 35 GHz radars than 95 GHz radars, primarily due to lower atmospheric losses at 35 GHz. Further information on MMW radars is given in [9].

In detecting obstacles, MMW radars are of particular value in locating poles and wires in fog. In rainy weather, however, because of rain backscattering effects, the detection of off-normal wires and wooden poles become very limited, as summarized in Table 4.11.

Table 4.11 Summary of 35 and 95 GHz MMW Radar Ranges (in km) for Detecting Wires and Poles

Condition	*Perpendicular to Wire (35 GHz/ 95 GHz)*	*Off-Normal to Wire (35 GHz/ 95 GHz)*	*Wooden Poles (35 GHz/ 95 GHz)*	*Conducting Poles (35 GHz/ 95 GHz)*
Clear	23/22	2.4/1.1	2.8/2.2	30/27
Fog	31/19	2.5/1.1	3.0/2.2	42/23
Light rain	21/10	None	None	27/13
Moderate rain	11/6	None	None	13/7
Heavy rain	4.5/3	None	None	5/3

REFERENCES

1. E. K. Reedy, N. T. Alexander, J. C. Butterworth, and J. A. Sheer, "High Power Coherent 95 GHz Radar System," *Proc. SPIE, MMW Technology II*, August 23–24, 1983.
2. S. A. Hovanessian, "Detection of Non-moving Targets by Airborne MMW Radars," *Microwave J.*, March 1986, pp. 159–166.
3. James C. Wiltse, "Millimeter Wave Trends," *Millimeter Wave Tech.*, Vol. 337, 1982.
4. S. A. Hovanessian, *Radar System Design and Analysis*, Artech House, Dedham, MA, 1984.
5. Kenneth J. Button, James C. Wiltse, eds., *Infrared and Millimeter Waves*, Volume 4, Academic Press, New York, 1981, pp. 46–47.
6. Stephen I. Johnston, ed., *Millimeter Wave Radar*, Artech House, Dedham, MA, 1980, pp. 137–165, 228, 625.
7. *Aviation Week and Space Tech.*, July 6, 1987, p. 103.
8. D. Van Vleck et al., "Theory of Radar Reflections from Wires or Thin Metallic Strips," *J. Applied Physics*, 1947, p. 274.
9. H. C. Currie, ed, *Principles and Applications of Millimeter Wave Radars* Artech House, Norwood, MA, 1988.

Chapter 5
Thermal Imaging Systems

Electro-optical thermal imagers, also referred to as *forward looking infrared* (FLIR) systems, and *infrared search and tracking* (IRST) systems are both passive devices which operate in the infrared wavelength region of the frequency spectrum. Thermal imaging systems use the temperature gradient of object scene to produce TV-like images at night as well as during the day. For this reason, they are also referred to as *night vision devices* (see also Chapter 1).

Infrared search and tracking systems are used primarily for detection of "hot spots" in large volume search and subsequent tracking of designated targets such as airborne and spaceborne targets against sky backgrounds. In the search mode of operation, they provide accurate angular target position information. In the tracking mode of operation, the IRST system can be locked onto a single target, tracking that target in angle. The IRST system can also be used in a track-while-scan mode, where several targets are tracked by a computer and the IRST system remains in the search mode updating target angular positions periodically.

The electro-optical system design and performance equations, together with numerical examples of these equations, are given in this chapter. Initially, the S/N ratio equation is derived for a point source target. Following this, we obtain the noise equivalent flux density relation. For distributed targets or background, the S/N ratio equation, including the integral of target spectral emittance, is given. The performance criteria depicted by noise equivalent temperature and minimum resolvable temperature are also derived and discussed. We give the probability of detection curves as a function of S/N ratio, with the probability of false alarm as a parameter, and use these curves in numerical examples. We also discuss image quality considerations relating target detection, orientation, recognition, and identification ranges to MRT values and illustrate them through numerical examples.

The IRST system design and performance equations are similar to those of thermal imaging systems for point targets against a uniform background. Because of this similarity, discussions of thermal imaging and IRST systems are combined in this chapter. In the first part of the chapter, we will discuss thermal imaging system design and performance equations, and show examples of these systems. The latter part of the chapter will cover IRST systems with particular attention to specifics of these systems, such as threshold settings and false alarm rate computations.

5.1 INTRODUCTION

The basic equations in the design and analysis of electro-optical thermal imaging systems, as well as IRST systems, involve target signal power and system noise power. As expected, the higher the S/N ratio, the greater is the probability of target detection at a given threshold. And, the higher the threshold, the lower is the number of false alarms.

For a point source target radiating in a narrow bandwidth with radiation intensity J_t, the irradiance at the entrance aperture of the optics is given by

$$H_t = \frac{J_t}{R^2} \tag{5.1}$$

where

J_t = radiation intensity, W/unit solid angle
H_t = irradiance at the optics aperture, W/cm^2
R = sensor-target range, cm

Note that Eq. (5.1) does not include any atmospheric transmission loss. Atmospheric losses, however, should be included in the final target signal values.

Multiplying Eq. (5.1) by the optical aperture area, we get the total input signal power:

$$S = \frac{\pi D^2}{4} \cdot \frac{J_t}{R^2} \tag{5.2}$$

where

S = signal power at the optical aperture, W
D = diameter of optical aperture, cm

The signal represented by Eq. (5.2) is at the entrance aperture of the optics; some optical transmission loss will occur between this point and the detector, as will be discussed later in the section on system losses.

The detector noise level is given by the relation

$$N = (A_d \Delta f)^{1/2}/D^* \tag{5.3}$$

where

N = noise power level, W
A_d = detector area, cm^2
Δf = frequency bandwidth of the received signal, Hz
D^* = detector detectivity, cm $Hz^{1/2}$/W

The terms of the noise power equation, (5.3), need additional discussion, as given in the following paragraphs.

For a square detector, the detector area, A_d, is related to the *instantaneous field of view* (IFOV), α_d, and the focal length, f_1, as follows and described in Figure 5.1:

$$\begin{aligned} \text{detector area} &= A_d \\ &= (\alpha_d^{1/2} f_1)(\alpha_d^{1/2} f_1) \\ &= \alpha_d f_1^2 \end{aligned} \tag{5.4}$$

The focal length is related to focal number, $f/\#$, and the aperture diameter by

$$f_1 = (f/\#)D \tag{5.5}$$

Using this, the detector area becomes

$$A_d = \alpha_d[(f/\#)D]^2 \tag{5.6}$$

To calculate the next term of the noise power equation, Δf, we need to calculate the detector on-target time, or equivalently, the time duration during which the detector receives target energy. As we will see later, this time duration can be converted to frequency increment Δf through Fourier transform analysis.

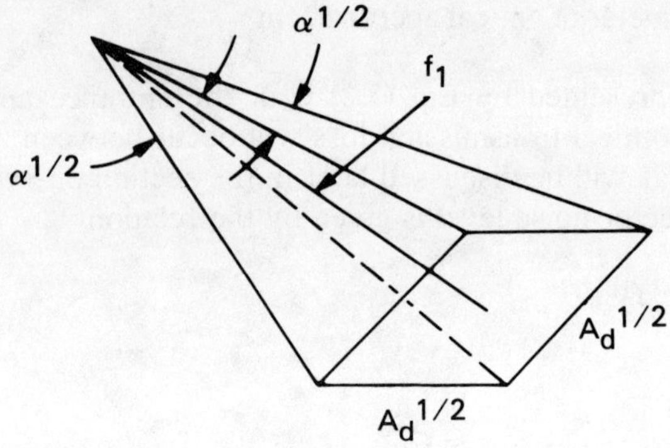

Figure 5.1 Detector area, focal length, and instantaneous field of view.

Assuming the total solid angle coverage of the sensor to be Ω_s sr, then the number of resolution elements will be this value divided by the IFOV of each detector:

$$n_r = \frac{\Omega_s}{\alpha_d} \tag{5.7}$$

where

n_r = number of resolution elements
Ω_s = total solid angle coverage of the sensor, sr
α_d = instantaneous field of view of each detector, sr

For a frame time of T_f, the on-target time per resolution cell will be

$$\begin{aligned} t_d &= T_f/n_r \\ &= \alpha_d T_f/\Omega_s \end{aligned} \tag{5.8}$$

where

T_f = frame time, s
t_d = on-target time, s

The value of the on-target time given by Eq. (5.8) is for a single detector. If the system is equipped with n detectors in parallel, the on-target time per detector will be increased by this factor:

$$t_d = \frac{n\alpha_d T_f}{\Omega_s} \tag{5.9}$$

where n is the number of detectors.

Note that Eq. (5.9) does not include scanning efficiency and other losses associated with detector on-target time. These losses will result in a scanning loss factor, which can be incorporated in Eq. (5.9) as follows:

$$t_d = \frac{n\alpha_d T_f}{\Omega_s} K_n$$

where K_n is the total scanning loss factor.

The relationship between on-target time and the bandwidth of the received signal is obtained from Fourier transform analysis. The Fourier transform of a step function of duration, t_d, can be expressed by a sin x/x relationship. This sin x/x represents the spectrum of the step function time signal. The approximate half-power bandwidth (3-dB) of this spectrum is $1/t_d$ with the null-to-null bandwidth of $2/t_d$ [1]. Using the fact that the filter bandwidth is inversely proportional to the on-target time with proportionality constant K_{fd}, we have

$$\Delta f = \frac{K_{fd}}{t_d} \tag{5.10}$$

where

Δf = equivalent filter noise bandwidth, Hz
K_{fd} = bandwidth proportionality constant

Using the value of t_d from Eq. (5.9) in (5.10) we get

$$\Delta f = \frac{\Omega_s K_{fd}}{n\alpha_d T_f K_n} \tag{5.11}$$

Substituting the parameters of the noise power equation, (5.3), from Eqs. (5.6) and (5.11), we get

$$N = \frac{D(f/\#)}{D^*}\left(\frac{\Omega_s}{nT_f}\cdot\frac{K_{fd}}{K_n}\right)^{1/2} \tag{5.12}$$

Using Eq. (5.12) together with the signal power equation, (5.2), we can write the S/N ratio equation as

$$\text{S/N} = \frac{\pi D J_t D^* L}{4(f/\#)R^2}\left(\frac{nT_f}{\Omega_s}\right)^{1/2} \tag{5.13}$$

The loss factor, L, has been added to the numerator of this equation to account for various system losses. This factor includes

$$L = K_o K_c K_d K_m (K_n/K_{fd})^{1/2}$$
$$= r_s(K_n/K_{fd})^{1/2} \tag{5.14}$$

where

K_o = optics
K_c = electronics
K_d = detector
K_m = monitor
r_s = product of these representing the system *modulation transfer function* (MTF)
K_n = total scanning loss factor
K_{fd} = bandwidth proportionality constant

The values of coefficients K given in Eq. (5.14) usually range between 0.5 and 1.0 with the exception of K_{fd}, which may be greater than unity. The coefficient K_{fd} may also be included in the system MTF value r_s.

Note that S/N Eq. (5.13) does not include atmospheric losses that are range dependent. These losses can be calculated by methods outlined at the end of this chapter.

Equation (5.13) represents the fundamental performance equation of electro-optical thermal imager and IRST systems, and gives the S/N ratio of a point source target at the detector. Using this equation, we can calculate the performance of an IRST system, in detection of ''hot spots'' against a clear background, as shown in the following example. Using (5.13), we can also develop a number of other parameters, such as minimum resolvable temperatures, which facilitates the performance computation of thermal imaging systems.

5.1.1 Example of an S/N Calculation

Consider an electro-optical IRST system used for the detection of point targets having the following system and target parameters:

$$\begin{aligned}
\text{radiation intensity} &= 1\ \text{kW} \\
\text{detector instantaneous field of view} &= 1\ \text{mr} \times 1\ \text{mr} \\
\text{sensor field of view} &= 20^\circ\ \text{Az} \times 5^\circ\ \text{El} \\
\text{sensor frame time} &= 5\ \text{s} \qquad (5.15) \\
\text{number of detectors} &= 10 \\
\text{optical diameter} &= 10\ \text{in} \\
\text{focal number } (f/\#) &= 1 \\
\text{detectivity} &= 10^{10}\ \text{cm Hz}^{1/2}/\text{W}
\end{aligned}$$

$$K_o = 0.5;\ K_e = 0.7;\ K_d = 0.84;\ K_m = 0.84;\ K_n = 1;\ K_{fd} = 1$$

Using these parameters, calculate the S/N at a range of 50 nmi. Assuming that a 90 percent single-look probability of detection is required for this target at this range, calculate the appropriate threshold setting (in dBs) and also the false alarm rate with this threshold.

Substituting Eq. (5.14) in (5.13) and using the parameters given in the problem, we have

$$\begin{aligned}
\text{S/N} &= \frac{\pi(10 \times 2.54)(10^3)(10^{10})(0.5)(0.7)(1)(0.84)(0.84)}{4(1)(50 \times 6080/3.28)^2(100)^2} \\
&\quad \times \{10 \times 5/[(20 \times 5)/(57.3)^2]\} \\
&= 23.05 \\
&= 13.62\ \text{dB} \qquad (5.16)
\end{aligned}$$

Conversion factors used in Eq. (5.16) include: 2.54 cm/in, 6080 ft/nmi, 3.28 ft/m, 100 cm/m, and 57.3 deg/rad.

From Figure 5.2, an S/N ratio of 13.6 dB and 90 percent probability of detection will result in a probability of false alarm of 10^{-7} (point *A* of the figure). The threshold setting for this false alarm probability is given approximately by the 50 percent probability of detection S/N as 12 dB (point *B* of Figure 5.2). (For a complete discussion of S/N and probability of detecting a sinusoidal signal in Gaussian noise, see [1].)

Using this probability of false alarm, we can compute the *false alarm rate* (FAR) as follows. During each on-target time, t_d, the set of detectors, n, is searched for the existence of a signal. If no signal is present in a detector, the probability of threshold crossing will be due to noise alone and the FAR will be given as

$$\text{FAR} = \frac{nP_n}{t_d} K_n \tag{5.17}$$

where

FAR = false alarm rate, false alarms/s
n = number of detectors
t_d = on-target time, s

Using Eq. (5.9) for t_d in Eq. (5.17), we get

$$\text{FAR} = \frac{P_n \Omega_s}{\alpha_d T_f} K_n \tag{5.18}$$

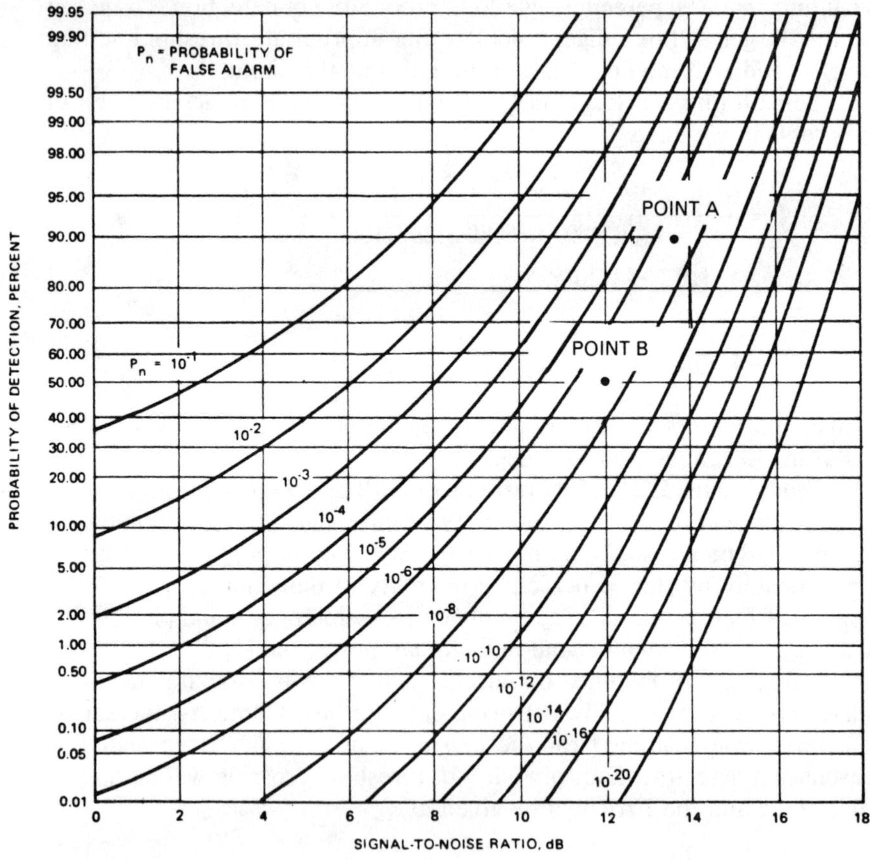

Figure 5.2 Probability of detection *versus* S/N ratio.

Using the parameters of the example problem, Eq. (5.18), results in

$$\text{FAR} = \frac{10^{-7}\left(\dfrac{20 \times 5}{57.3^2}\right)}{(10^{-3})^2\,(5)}$$
$$= 6.09 \times 10^{-4} \text{ false alarms/s}$$
$$= 2.19 \text{ false alarms/h} \tag{5.19}$$

5.2 NOISE EQUIVALENT FLUX DENSITY

Another often-used system performance measure is the *noise equivalent flux density* (NEFD). This value is defined as irradiance at the optics per unit S/N ratio, or

$$\text{NEFD} = \frac{H_t}{\text{S/N}} \tag{5.20}$$

Using Eqs. (5.1) and (5.13) in (5.20), we get

$$\text{NEFD} = \frac{4(f/\#)}{\pi D D^* L}\left(\frac{\Omega_s}{nT_f}\right)^{1/2} \tag{5.21}$$

The S/N and the NEFD equations are used to predict performance of existing electro-optical systems or design of new systems.

5.3 SPECTRAL EMITTANCE OF A TARGET

In the previous derivation of S/N ratio performance equations for point targets, we assumed that the background was clear (i.e., was free of spectral radiation emittance.) An example of this is an airborne target against a clear sky background. In a number of applications, however, targets of interest are located in an environment with considerable spectral radiation emittance, for example, a tank or a truck on a roadway or grass-covered field. In these cases the radiation emittance of the background increases the noise level of the system. Detection of the target should occur in the presence of this additional noise.

In the performance analysis of thermal imager systems used in detection of ground targets, we should take into account this background emittance. In the following analysis, we will derive thermal imager performance equations and include background emittance and target temperature gradient with respect to this background in our calculations. Therefore, the spectral emittance of a thermal target may be written

$$W(\lambda, T) = \epsilon(\lambda) W_b(\lambda, T) \tag{5.22}$$

where

$W(\lambda, T)$ = target spectral emittance, W/cm² sr μm
$\epsilon(\lambda)$ = target emissivity
$W_b(\lambda, T)$ = blackbody spectral emittance, W/cm² sr μm
T = target temperature, K
λ = wavelength, μm

The spectral emittance $W(\lambda, T)$, usually used for distributed thermal targets, is obtained by measurements as shown in Figure 5.3. This figure gives the spectral emittance of a grass-covered field at two different temperatures, 32°C and 10°C. The spectral emittance, using Lambertian radiation distribution, can be applied in the calculation of the target signal power as shown:

$$S = \frac{\pi D^2}{4} \alpha_d \int_{\lambda_1}^{\lambda_2} W(\lambda, T) d\lambda \tag{5.23}$$

where D is the aperture diameter and α_d, as before, is the IFOV. The units of the signal S of Eq. (5.23) will be in W. The integral is taken over the range of detector wavelengths.

Using Eq. (5.23), together with the noise power equation, (5.12), we can obtain the S/N ratio for a distributed target:

$$\text{S/N} = \frac{\pi D \alpha_d L \tau_a}{4(f/\#)} \left(\frac{n T_f}{\Omega_s}\right)^{1/2} \int_{\lambda_1}^{\lambda_2} W(\lambda, T) D^*(\lambda)\, d\lambda \tag{5.24}$$

The coefficient τ_a, denoting the optical transmission efficiency, is introduced in Eq. (5.24).

Using the spectral emittance curves of Figure 5.3, the differential spectral emittance $\partial W(\lambda)/\partial T$ can be calculated and plotted as shown. This curve is used in another system performance measure, the *noise equivalent temperature* (NET), as will be seen shortly.

The incremental target emittance can be obtained by using the differential of Eq. (5.22) for a given wavelength, λ:

$$dW(\lambda, T) = \frac{\partial W(\lambda, T)}{\partial \epsilon} d\epsilon + \frac{\partial W(\lambda, T)}{\partial T} dT \tag{5.25}$$

Using the difference notation, this becomes

$$\Delta W(\lambda, T) = \frac{\partial W(\lambda, T)}{\partial \epsilon} \Delta\epsilon + \frac{\partial W(\lambda, T)}{\partial T} \Delta T \tag{5.26}$$

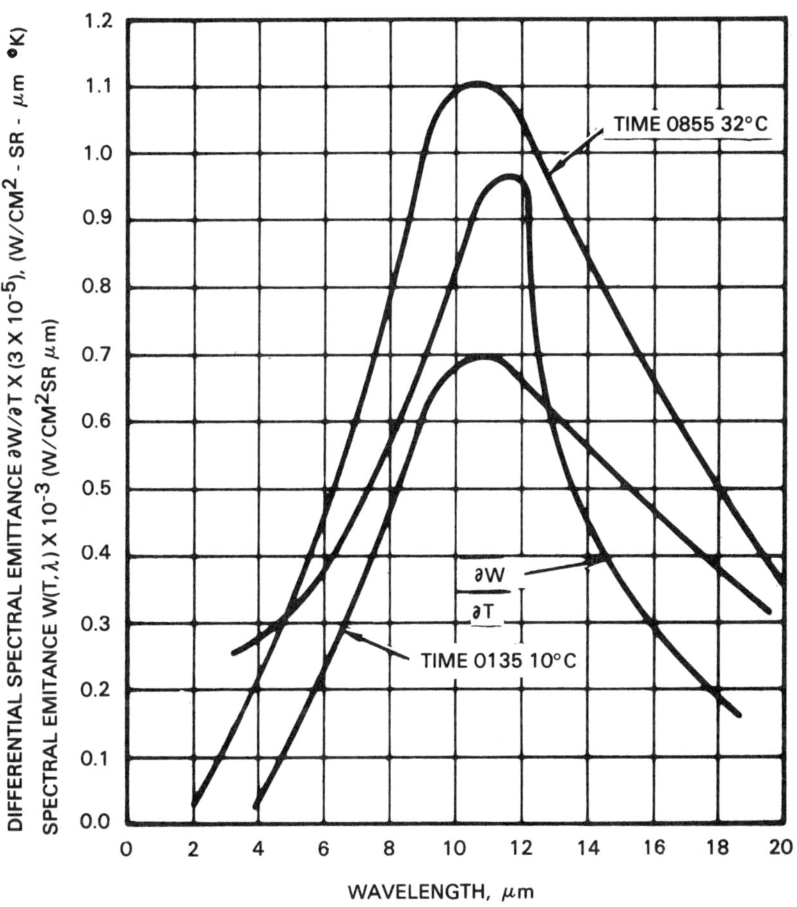

Figure 5.3 Day and night spectral emittance of grass covered field (from [2], pages 3–48).

Assuming that the background emissivity remains constant ($\Delta\epsilon = 0$), then the spectral emittance will change only with temperature as follows:

$$\Delta W(\lambda, T) = \frac{\partial W(\lambda, T)}{\partial T} \Delta T \tag{5.27}$$

Using this, the change in signal power due to an incremental temperature change, ΔT, can be computed using the signal power equation:

$$\Delta S = \frac{\pi D^2}{4} \alpha_d \int_{\lambda_1}^{\lambda_2} \frac{\partial W(\lambda, T)}{\partial T} \Delta T \, d\lambda \tag{5.28}$$

where ΔS is the incremental signal change due to temperature change of ΔT. Equation (5.28) can also be obtained directly by differentiating the signal power equation, (5.23), with respect to T.

Using the noise power equation, (5.12), the target signal power equation, (5.28), and the optical transmission efficiency, τ_a, we obtain

$$\Delta S/N = \frac{\pi D \alpha_d \tau_a D^* L}{4(f/\#)} \left(\frac{nT_f}{\Omega_s}\right)^{1/2} \int_{\lambda_1}^{\lambda_2} \frac{\partial W}{\partial T} \Delta T d\lambda \tag{5.29}$$

where the loss term L is incorporated in the numerator of the above equation.

5.4 NOISE EQUIVALENT TEMPERATURE

Another measure of optical system performance is depicted by the *noise equivalent temperature* (NET) and can be derived easily using the $\Delta S/N$ equation, (5.29). Assuming the detectivity, D^*, to be a function of wavelength, Eq. (5.29) can be solved for ΔT as a function of $\Delta S/N$. The value of ΔT thus obtained for a $\Delta S/N$ of unity is termed noise equivalent temperature:

$$\text{NET} = \frac{4(f/\#)}{\pi \tau_a D \alpha_d L} \left(\frac{\Omega_s}{nT_f}\right)^{1/2} \frac{1}{\displaystyle\int_{\lambda 1}^{\lambda_2} \frac{\partial W(\lambda, T)}{\partial T} D^*(\lambda) d\lambda} \tag{5.30}$$

Equation (5.30) can also be used as a design tool to establish requirements of electro-optical systems, as shown in the following example.

Consider the design of an IR imaging system onboard a spacecraft at 500 mi altitude and having a velocity of 4 mi/s [3]. The azimuth coverage is 12° and the elevation coverage is that obtained for 1 s space flight. It is desired to obtain a noise equivalent temperature of 0.10 K. Consider the following design parameters obtained after several iterations:

$$\begin{aligned} \text{diameter of the optics} &= 8 \text{ in} \\ \text{number of detectors} &= 50 \\ \text{spatial resolution} &= 200 \text{ ft} \\ f/\# &= 1 \\ \text{number of frames/s} &= 30 \\ \text{peak detectivity} &= 2 \times 10^{11} \text{ cm Hz}^{1/2}/W \end{aligned} \tag{5.31}$$

$$\tau_a = 0.7;\ K_o = 0.6;\ K_d = 0.8;\ K_m = 0.8;\ K_c = 0.8;$$
$$K_n = 0.81;\ K_{fd} = 1.5$$

Additionally, the normalized differential emittance and detector detectivity curves of Figure 5.4 are also given. Assume detectors are operating in the wavelength region of 8–14 μm.

Using this data, we can calculate a number of additional parameters to be used in conjunction with the NET equation, (5.30). From spacecraft velocity and altitude, we can calculate the equivalent elevation coverage for 1 s flight duration:

$$\begin{aligned} \epsilon &= \frac{Vt}{h} \\ &= \frac{4(1)}{500} = 8 \times 10^{-3} \text{ rad} \end{aligned} \tag{5.32}$$

Using this and the azimuth coverage of 12°, we obtain the spatial window:

$$\Omega_s = \frac{12}{57.3}(8 \times 10^{-3}) = 1.675 \times 10^{-3} \text{ sr} \tag{5.33}$$

The required resolution of 200 ft, will give the instantaneous detector field of view:

$$\alpha_d^{1/2} = \frac{200}{(500)(5280)} = 7.575 \times 10^{-5} \text{ rad} \tag{5.34}$$

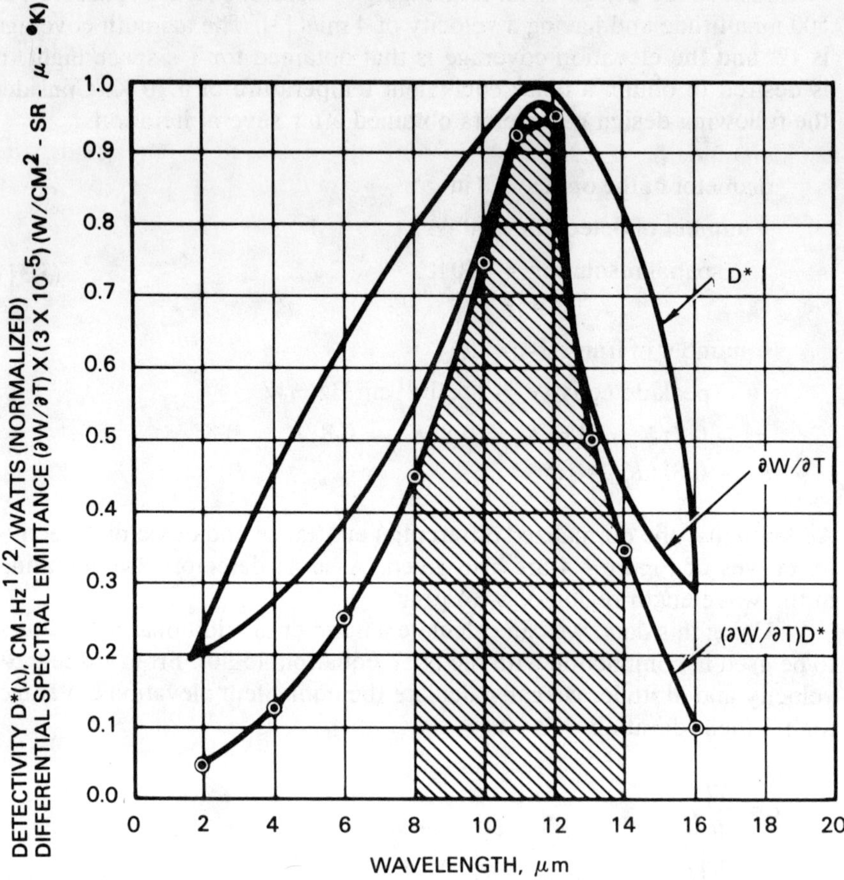

Figure 5.4 Spectral emittance and detectivity *versus* wavelength.

where 5280 is conversion from mi to ft. Using Eq. (5.34), α_d becomes

$$\alpha_d = 5.74 \times 10^{-9} \text{ sr} \tag{5.35}$$

Using these values, together with $L = 0.225$ computed from Eq. (5.14), in Eq. (5.30), we can write

$$\text{NET} = \frac{4(1)}{\pi(0.7)(8 \times 2.54)(5.74 \times 10^{-9})(0.225)} \times \left(\frac{1.675 \times 10^{-3}}{50(1/30)}\right)^{1/2}$$

$$\times \left(\frac{1}{(2 \times 10^{11})(1.2 \times 10^{-4})}\right)$$
$$= 0.0916\ \text{K} \tag{5.36}$$

In this equation, 1.2×10^{-4} represents the value of the integral in the denominator of Eq. (5.30) obtained by calculating the area under the curve of Figure 5.4 (dashed region). The multiplier of this value is the normalization constant $D^*(\lambda)$ given in Eq. (5.31).

The NET value of Eq. (5.36) is reasonably close to the design goal of 0.10 K.

From the given parameters, a number of other system constants can also be derived. For example, the detector on-target time from Eq. (5.9) becomes

$$\begin{aligned} t_d &= \frac{n\alpha_d T_f}{\Omega_s} \\ &= \frac{(50)(5.74 \times 10^{-9})(1/30)}{1.67 \times 10^{-3}} \\ &= 5.711 \times 10^{-6}\ \text{s} \end{aligned} \tag{5.37}$$

Using this value in Eq. (5.10) with noise bandwidth factor of 1.5, we get

$$\begin{aligned} \Delta f &= \frac{1.5}{t_d} \\ &= 262\ \text{kHz} \end{aligned} \tag{5.38}$$

representing the bandwidth of the received signal.

5.5 MINIMUM RESOLVABLE TEMPERATURE

Minimum resolvable temperature (MRT) can be derived from the NET by considering the display and observer parameters. Assuming a target resolution distance, T_x, from an observation point at a distance, R, away (see Figure 5.5), we get

$$\theta_d^{1/2} = \frac{1000\ T_x}{R} \tag{5.39}$$

where

$\theta_d^{1/2}$ = angular resolution, mr
T_x = linear resolution distance at the target, m
R = distance to observation point, m

The one-half power notation is used in Eq. (5.39) to conform with the previous IFOV notation of $\alpha^{1/2}$. The one-half power designation refers to a linear angular measurement, while the unity power designation refers to a two-dimensional measurement. Using the angular resolution of Eq. (5.39), we can define a spatial frequency as

$$f_s = \frac{1}{2\theta_d^{1/2}} \tag{5.40}$$

where f_s is the spatial frequency in c/mr.

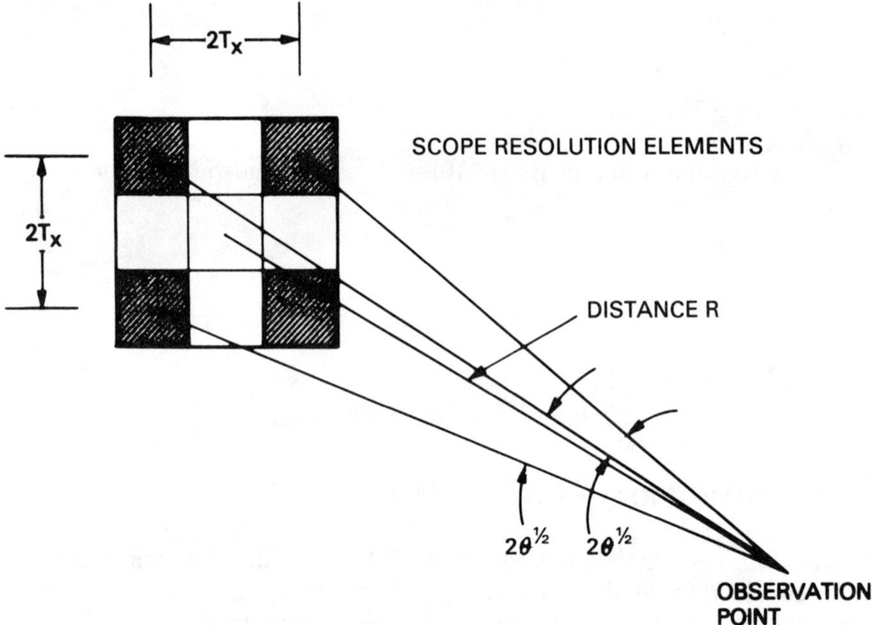

Figure 5.5 Target resolution elements and subtended angles.

For a square display cell, the subtended angle will be $\theta_d^{1/2}$ mr. This spatial frequency through Eqs. (5.39) and (5.40) is related to thermal imager target distance, R. Since the NET of the target is calculated per detector resolution cell, the equivalent display NET (referred to as MRT) will be modified by $\theta_d^{1/2}$ as follows:

$$\text{MRT} = (\text{NET})(\alpha_d/\theta_d)^{1/2} \tag{5.41}$$

where

$\alpha_d^{1/2}$ = detector field of view
$\theta_d^{1/2}$ = display resolution cell

Another factor affecting the MRT is the observer integration of the detected signal. Noting that the NET value was calculated for a single detector frame, this value should be modified by the number of frames integrated by the observer. The number of integrated frames can be written:

$$n_i = T_e\dot{F} \tag{5.42}$$

where

n_i = number of frames integrated by the observer
T_e = observer integration time, s
$\dot{F}$ = frame rate, frames/s

Using Eq. (5.42) in (5.41), we get

$$\text{MRT} = (\text{NET})\left(\frac{\alpha_d}{\theta_d}\right)^{1/2}\left(\frac{1}{T_e\dot{F}}\right)^{1/2} \tag{5.43}$$

Since detection of a signal in the display scope is probabilistic, a percent probability of detection will be associated with each S/N ratio or NET value. From Reference [4], page 146, a factor of about 3 is required for a 50 percent probability of detection. This assumes that, on the average, about 50 percent of resolution cells covering the target will be sufficient for target imaging. Incorporating this factor in Eq. (5.43) and using the spatial frequency relation, Eq. (5.40), we get

$$\text{MRT} = \frac{6\,(\text{NET})f_s(\alpha_d)^{1/2}}{(T_e\dot{F})^{1/2}} \tag{5.44}$$

Replacing the NET of this equation by its equivalent expression from Eq. (5.30), and using Eq. (5.14) for L, we get

$$\mathrm{MRT} = \frac{24(f/\#)}{\pi\tau_a D}\left(\frac{f_s}{r_s}\right)\left(\frac{K_{fd}}{K_n}\right)^{1/2} \times \left(\frac{\Omega_s}{nT_f\alpha_d}\right)^{1/2}\left(\frac{1}{T_e\dot{F}}\right)^{1/2}$$

$$\times \frac{1}{\displaystyle\int_{\lambda_1}^{\lambda_2} \frac{\partial W(\lambda, T)}{\partial T} D^*(\lambda)\, d\lambda} \tag{5.45}$$

This expression for MRT includes optical, detector, electronics, and monitor losses as these values are included in the r_s (modulation transfer function MTF) term. The overall system MTF, as discussed at length in Chapter 3 of [4], can be computed as a function of the spatial frequency of Eq. (5.40).

An example of MTF curve is given in [4], page 113, and reproduced here as Figure 5.6. The MTF curve of Figure 5.6 is normalized with respect to the limiting spatial frequency f_{ls} defined as follows:

$$\bar{f}_s = f_s/f_{ls} \tag{5.46}$$

where

$\bar{f}_s$ = normalized spatial frequency
f_s = spatial frequency
f_{ls} = limiting spatial frequency

The value of f_{ls} can be defined in terms of optical aperture (for optics limited designs) and the wavelength as

$$f_{ls} = \frac{D}{\lambda} \tag{5.47}$$

where

f_{ls} = limiting spatial frequency
D = optical aperture
λ = wavelength

The limiting spatial frequency, f_{ls}, should be obtained from detector IFOV for detector limited designs. The value of MRT obtained from Eq. (5.45)

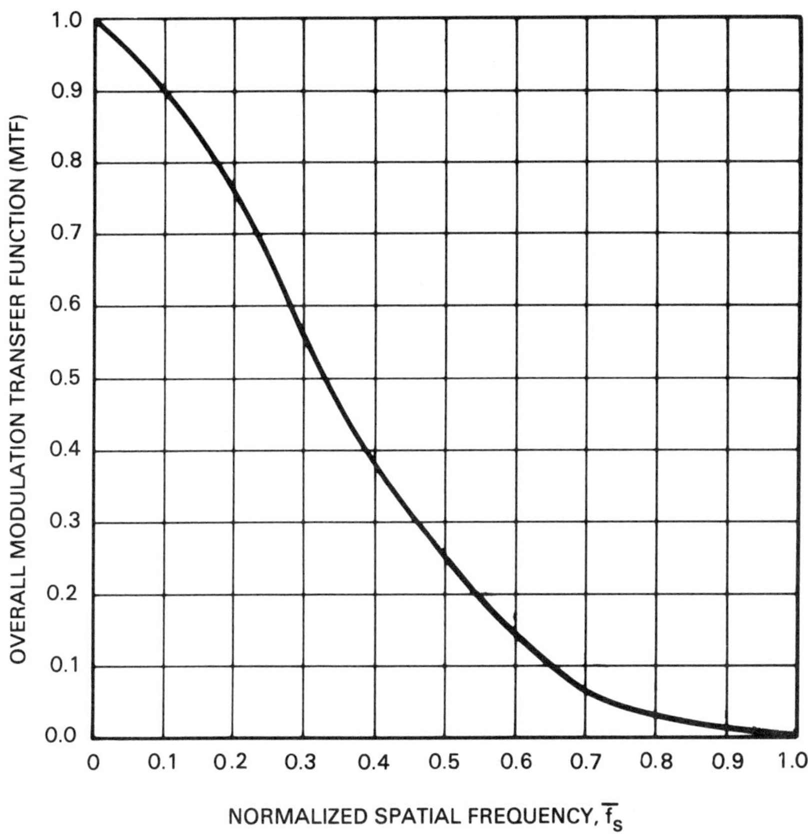

Figure 5.6 Normalized MTF *versus* normalized spatial frequency, $\bar{f}_s$.

is another important parameter used in performance and design analysis of electro-optical systems.

5.6 IMAGE QUALITY CONSIDERATIONS

Through experiments with observers, Johnson [5] determined the number of line pairs required to detect, determine orientation, recognize, and identify tactical targets. These results are illustrated in Figure 5.7. It can be seen from the table accompanying the figure that the requirements in each category are relatively independent of the target. Recognition, which is the most frequently used figure of merit for system performance

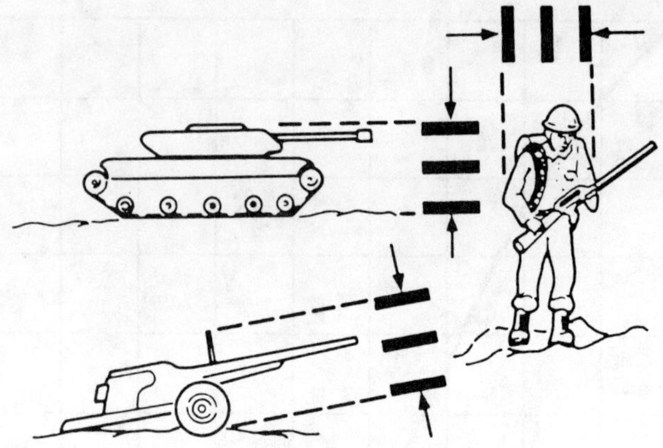

Target	Resolution per minimum dimension in line pairs			
Broadside view	Detection	Orientation	Recognition	Identification
Truck	0.90	1.25	4.5	8.0
M-48 tank	0.75	1.20	3.5	7.0
Stalin tank	0.75	1.20	3.3	6.0
Centurion tank	0.75	1.20	3.5	6.0
Half-track	1.00	1.50	4.0	5.0
Jeep	1.20	1.50	4.5	5.5
Command car	1.20	1.50	4.3	5.5
Soldier (standing)	1.50	1.80	3.8	8.0
105 Howitzer	1.00	1.50	4.3	6.0
Average	1.0 ± 0.25	1.4 ± 0.35	4.0 ± 0.8	6.4 ± 1.5

Figure 5.7 Required resolution for detection, orientation, recognition, and identification (from [5]).

calculations, requires four line pairs across the minimum target dimension. On the other hand, simple target detection requires only one line pair across the minimum target dimension. The table of Figure 5.7 also gives the average values of line pairs (cycles) across the minimum target dimension for detection, orientation, recognition, and identification.

5.7 NUMERICAL EXAMPLE OF MRT CALCUATION

As a numerical example of MRT calculations, consider a thermal imaging system with the following parameters:

spectral band	$= 8\text{–}12\ \mu\text{m}$
optical diameter	$D = 8$ in
focal number	$(f/\#) = 2$
filter bandwidth factor	$K_{fd} = 1$
scanning efficiency	$K_n = 0.8$
sensor field of view	$= 2.0°(\text{Az}),\ 1.0°(\text{El})$
number of detectors	$n = 1024$
frame time	$T_f = 3.34 \times 10^{-2}$ s
detector instantaneous field of view	$\alpha_d^{1/2} = 50\ \mu\text{rad}$
frame rate	$\dot{F} = 30/\text{s}$
eye integration time	$T_e = 0.2$ s
peak detectivity	$D^* = 2 \times 10^{11}$ cm Hz$^{1/2}$/W
optical efficiency	$\tau_a = 0.5$

(5.48)

In addition to these parameters, use the spectral emittance and detectivity characteristics of Figure 5.4. Using this figure:

$$\int_8^{12} \frac{\partial W[\lambda, T]}{\partial T} D^*(\lambda)\, d\lambda = 8.4 \times 10^{-5}\ \text{Hz}^{1/2}/\text{cm sr K} \tag{5.49}$$

Note that the integral of Eq. (5.49) is taken between the wavelength limits of 8 and 12 μm rather than 8 and 14 μm, shown by the dashed area of Figure 5.4. Use an atmospheric transmission coefficient of unity and calculate the atmospheric losses as discussed in the next section.

Using the parameters of Eq. (5.48), calculate the detection and recognition range of the system for a tank-type, 3 m × 3 m target with a temperature difference ΔT of 2.5°. Using these system parameters, we can calculate the parameters of MRT equation (5.45):

$$\begin{aligned} \Omega &= \frac{2 \times 1}{(57.3)^2} \\ &= 6.09 \times 10^{-4}\ \text{sr} \\ \alpha_d &= (50 \times 10^{-6})^2 \\ &= 2.5 \times 10^{-9}\ \text{sr} \end{aligned} \tag{5.50}$$

Using the given parameters in Eq. (5.45), we get

$$\begin{aligned}
\text{MRT} &= \frac{(24)(2)}{\pi(8 \times 2.54)}\left(\frac{f_s}{r_s}\right)\left(\frac{1}{0.8}\right)^{1/2} \\
&\times \left(\frac{6.09 \times 10^{-4}}{(1024)(3.34 \times 10^{-2})(2.5 \times 10^{-9})}\right)^{1/2} \\
&\times \left(\frac{1}{0.2 \times 30}\right)^{1/2}\left(\frac{1}{(8.4 \times 10^{-5})(2 \times 10^{11})}\right) \\
&= 1.724\text{E} - 6\left(\frac{f_s}{r_s}\right) \quad (f_s \text{ in c/rad}) \\
&= 1.724\text{E} - 3\left(\frac{f_s}{r_s}\right) \quad (f_s \text{ in c/mr})
\end{aligned} \tag{5.51}$$

Using Eq. (5.51), with a limiting spatial frequency of

$$\begin{aligned}
f_{ls} &= \frac{D}{\lambda} \\
&= \frac{8 \times 2.54 \times 10^{-2}}{10 \times 10^{-6}} \\
&= 20.32 \text{ c/mr}
\end{aligned} \tag{5.52}$$

together with the r_s values of Figure 5.6, we get the spatial frequency f_s and the corresponding MRT values found in Table 5.1. The values of Table 5.1 are plotted on Figure 5.8, showing the exponential nature of the MRT curve.

Noting that the target has a dimension of 3 m, each spatial frequency cycle will represent a range of 3 km; the reason being that each cycle represents a one mr subtended angle.

From Figure 5.8, the given tank ΔT of 2.5° will correspond to a spatial frequency of 17.6 c/mr. From Figure 5.7, on the average one line pair or one cycle is required for *detection*. Using this, the detection range of the thermal sensing system becomes

$$(3 \text{ km})\,(17.6 \text{ c}) = 52.8 \text{ km} \tag{5.53}$$

From Figure 5.7, the average number of *recognition* line pairs is 4. Using this, the recognition range becomes

$$(3)(17.6/4) = 13.2 \text{ km} \tag{5.54}$$

Note that the detection and recognition values given in Eqs. (5.53) and (5.54) do not include atmospheric losses, which will considerably reduce these ranges, as will be seen in the next section.

Table 5.1 Calculated Values for Spatial Frequency and MRT

Spatial Frequency (f_s)	*MRT (K)*	*Spatial Frequency* (f_s)	*MRT (K)*
1	0.00182	11	0.08225
2	0.00384	12	0.12918
3	0.00618	13	0.22421
4	0.00909	14	0.40287
5	0.01287	15	0.64810
6	0.01812	16	0.92253
7	0.02335	17	1.47138
8	0.03253	18	3.11792
9	0.04404	19	—
10	0.05919	20	—

5.8 ATMOSPHERIC LOSS CONSIDERATIONS

In a report by Obert et al. [6], atmospheric transmittance of thermal imaging systems in the 8–12 μm region is given as a function of range and other parameters including visibility range, air temperature, and relative humidity. This table is reproduced here as Table 5.2.

Selected conditions from this table were converted to decibel units and plotted to obtain transmittance values per km, as shown in Figures 5.9 and 5.10. Figure 5.9 represents transmittance values for clear weather with visibility greater than 4 km and Figure 5.10 represents conditions of 1 km visibility under fog, with air temperatures of less than and greater than 10° C.

From Figures 5.9 and 5.10, we see that the atmospheric loss *versus* range data points represent constant slope lines; that is, atmospheric transmittance in dB/km remains constant. The transmittance values of Figure 5.9 are summarized in Table 5.3. As we see in this table, the atmospheric transmittance for clear air with a visibility greater than 4 km varies between 0.72 and 0.97, depending on relative humidity and air temperature.

From Figure 5.10, the atmospheric attenuation for fog, with 1 km visibility, ranges between 3 and 4 dB/km, depending on the air temperature:

air temperature >10°C 4 dB/km
(a factor of 0.40)

air temperature <10°C 3 dB/km
(a factor of 0.50)

In a similar fashion, atmospheric transmittance in dB/km can be computed for snow and rain, using values from Table 5.2. For snow, the atmospheric transmittance will range from 7 to 17 dB/km, depending on visibility and temperature. In rain, the atmospheric transmittance will vary between 4 and 17 dB/km, depending on the values of the same visibility and temperature.

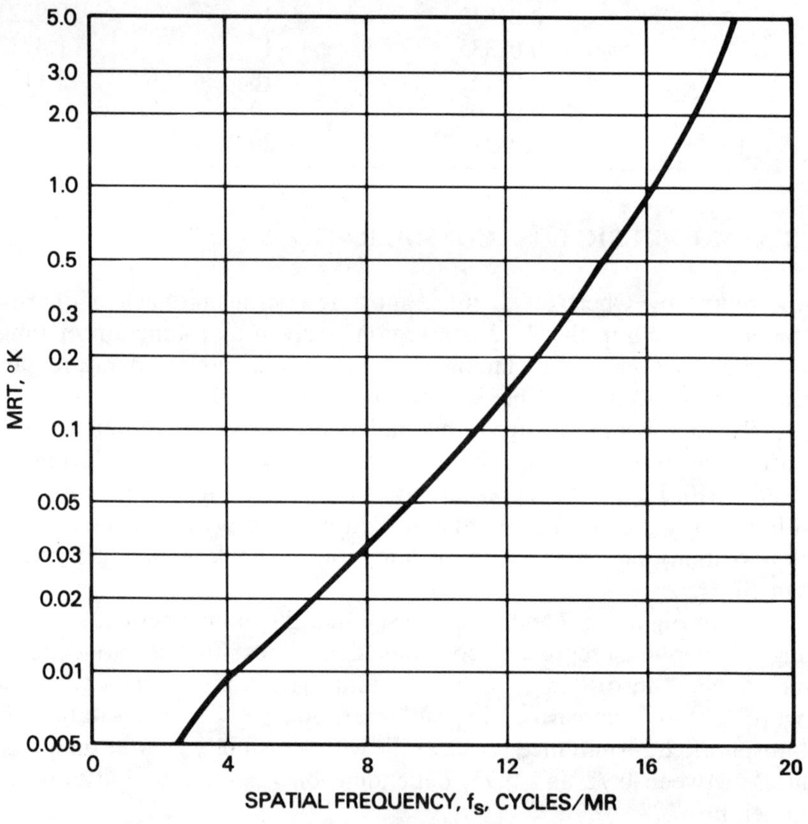

Figure 5.8 MRT *versus* spatial frequency for numerical example.

Table 5.2 Atmospheric Transmittance for an 8–12 Thermal Imaging System (from [6])

WEATHER OBSERVABLES			TRANSMISSION (%) FOR TARGET RANGE FROM 0.5 - 15.0 KILOMETERS																	
V.R.(km)	T_{AIR}(°C)	R.H.(%)	0.5	.75	1.0	1.5	2.0	2.5	3.0	3.5	4.0	4.5	5.0	6.0	7.0	8.0	9.0	10.0	12.5	15.0
FOG																				
0.25	All	All	0	0	0	0	0	0	0	0	0	0	0	0	0	0	0	0	0	0
0.25 - 0.35	All	All	5	1	0	0	0	0	0	0	0	0	0	0	0	0	0	0	0	0
0.35 - 0.45	All	All	16	6	3	0	0	0	0	0	0	0	0	0	0	0	0	0	0	0
0.45 - 0.70	≥10	All	38	24	15	6	2	1	0	0	0	0	0	0	0	0	0	0	0	0
0.45 - 0.70	<10	All	43	29	19	9	4	2	1	0	0	0	0	0	0	0	0	0	0	0
0.70 - 0.90	≥10	All	53	39	29	16	9	5	3	2	1	1	0	0	0	0	0	0	0	0
0.70 - 0.90	<10	All	60	47	37	23	14	9	5	3	2	1	1	0	0	0	0	0	0	0
0.90 - 1.25	≥10	All	62	49	40	26	17	11	7	5	3	2	1	0	0	0	0	0	0	0
0.90 - 1.25	<10	All	70	59	50	36	26	19	13	10	7	5	4	2	1	1	0	0	0	0
1.25 - 1.75	≥10	All	72	62	53	40	30	23	17	13	10	8	6	3	2	1	1	0	0	0
1.25 - 1.75	<10	All	81	73	67	56	46	39	33	27	23	19	16	12	8	6	4	3	1	1
1.75 - 2.25	≥10	All	75	66	59	46	36	29	23	18	15	21	10	6	4	3	2	1	0	0
1.75 - 2.25	<10	All	85	79	74	64	56	50	44	39	34	30	27	21	16	13	10	8	5	3
2.25 - 3.00	≥10	All	77	68	61	49	40	32	26	21	18	14	12	8	6	4	3	2	1	0
2.25 - 3.00	<10	All	87	82	77	69	61	55	50	45	40	36	33	27	22	18	15	12	8	5
3.00 - 4.00	≥10	All	78	70	62	51	41	34	28	23	19	16	13	10	6	5	3	2	1	0
3.00 - 4.00	<10	All	88	83	79	71	64	58	53	48	44	40	37	31	26	22	18	15	10	7
CLEAR																				
>4.00	>10	80-100	82		69	59	50	43	37	32	28	24	21	16	12	9	7	6	3	1
>4.00	>10	50-80	87		77	69	63	57	51	46	42	38	35	29	24	20	17	14	9	5
>4.00	>10	<50	96		93	91	89	87	85	84	82	81	79	76	74	72	72	69	63	57
>4.00	<10	>50	94		89	86	82	79	76	73	71	68	66	62	58	54	51	48	40	33
>4.00	<10	<50	98		96	95	93	92	91	90	89	88	87	86	85	83	82	81	77	73
SNOW																				
1.00	≤10	All	6		0	0	0	0	0	0	0	0	0	0	0	0	0	0	0	0
1.50	≤10	All	15		2	0	0	0	0	0	0	0	0	0	0	0	0	0	0	0
2.00	≤10	All	24		6	1	0	0	0	0	0	0	0	0	0	0	0	0	0	0
3.00	≤10	All	38		15	6	1	0	0	0	0	0	0	0	0	0	0	0	0	0
4.00	≤10	All	47		20	11	5	3	1	0	0	0	0	0	0	0	0	0	0	0
RAIN																				
1.00	All	All	5		0	0	0	0	0	0	0	0	0	0	0	0	0	0	0	0
1.50	All	All	17		2	0	0	0	0	0	0	0	0	0	0	0	0	0	0	0
2.00	>10	All	25		7	2	0	0	0	0	0	0	0	0	0	0	0	0	0	0
2.00	≤10	All	28		8	2	1	0	0	0	0	0	0	0	0	0	0	0	0	0
3.00	>10	All	35		13	5	2	1	0	0	0	0	0	0	0	0	0	0	0	0
3.00	≤10	All	44		18	8	3	1	0	0	0	0	0	0	0	0	0	0	0	0
4.00	>10	<50	42		26	13	7	3	1	0	0	0	0	0	0	0	0	0	0	0
4.00	≤10	<50	52		32	18	10	6	3	2	1	0	0	0	0	0	0	0	0	0
5.00	>10	<50	54		31	18	10	6	4	2	1	0	0	0	0	0	0	0	0	0
5.00	≤10	<50	62		38	24	15	10	6	4	2	0	0	0	0	0	0	0	0	0

For the purposes of illustration, consider the second line of the *clear case* of Table 5.2. In this case, visibility is greater than 4 km, the air temperature is greater than 10°, and the relative humidity ranges between 50 and 80 percent. From Table 5.3, this corresponds to an atmospheric attenuation loss of 0.9 dB/km.

Table 5.3 Atmospheric Transmittance for Clear Weather with Visibility Greater Than 4 km

Temperature	*>10°C*			*<10°C*	
Relative Humidity	*<50%*	*>50%*	*80–100%*	*<50%*	*>50%*
Atmospheric attenuation Factor/km	0.95 (0.2 dB)	0.81 (0.9 dB)	0.72 (1.4 dB)	0.97 (0.1 dB)	0.92 (0.35 dB)

The trial-and-error calculations resulting in the detection ranges that include atmospheric losses are given in Table 5.4. This table includes target sensor range and the corresponding atmospheric transmission loss calculated by using 0.9 dB/km loss factor. The value of $\overline{\Delta T}$, which includes the transmission loss, is obtained by multiplying the design ΔT of 2.5° by the transmission loss factor. Using this $\overline{\Delta T}$ value and the curve of Figure 5.8 (or a computer printout), the new value of spatial frequency is obtained. Multiplying this value by 3 km will result in the detection range with atmospheric losses.

Table 5.4 Calculation of Atmospheric Loss and Corresponding Detection Range

Target Sensor Range (km)	*Transmission Loss Factor (τ_a)*	$\overline{\Delta T} = \Delta T$ $x(\tau_a)$	*Spatial Frequency*	*New Range (km)*
10	0.126	0.315	13.6	40.8
11	0.102	0.255	13.2	39.6
12	0.083	0.207	12.8	38.4
13	0.068	0.170	12.4	37.2
14	0.054	0.135	12.0	36.0
15	0.045	0.112	11.8	35.4
16	0.036	0.090	11.2	33.6
17	0.029	0.072	10.5	31.5
18	0.024	0.060	9.95	29.85
19	0.019	0.047	9.14	27.42
20	0.016	0.040	8.53	25.59
21	0.013	0.032	7.92	23.76
22	0.010	0.025	7.10	21.30
23	0.008	0.020	6.50	19.50
24	0.007	0.017	5.90	17.70
25	0.006	0.015	5.50	16.50

From Table 5.4, we see that there is a possible match between the starting range and final target sensor range of about 21 km. Thus, the detection range of the thermal sensor under the stated atmospheric condition is approximately 21 km as opposed to the *no-loss range* of 52.8 km.

In a similar manner, recognition ranges that include atmospheric losses can be calculated as shown in Table 5.5. Since the no-loss recognition range from Eq. (5.54) was about 13 km, target sensor ranges of Table 5.4 were taken around this value. The atmospheric transmission losses were selected from the second *clear case* of Table 5.2 and a four-cycle recognition criterion was used in computing the new range values of this table. From Table 5.5, we see that a match is obtained at a range of about 10 km. Thus, the recognition range of thermal sensor will decrease from its value of about 13 km given in Eq. (5.54) to 10 km, which includes atmospheric transmission losses.

Table Table 5.5 Calculation of Atmospheric Loss and the Corresponding Recognition Range

Target Sensor Range (km)	*Transmission Loss Factor* (τ_a)	$\overline{\Delta T} = \Delta T$ $x(\tau_a)$	*Spatial Frequency*	*New Range (km)*
9	0.155	0.387	14.0	10.50
10	0.126	0.315	13.6	10.20
11	0.102	0.255	13.2	9.90
12	0.083	0.207	12.8	9.60
13	0.068	0.170	12.4	9.30
14	0.054	0.135	12.1	9.07
15	0.045	0.112	11.7	8.78

5.9 COMPUTER PROGRAM

A computer program was developed to calculate performance values. (This section of the chapter and the corresponding computer program was written by Lori Bean.) The program, written in BASIC for use on the IBM PC, utilizes graphics capabilities to "box in" its input and output displays. It finds MRT values corresponding to spatial frequencies and determines target detection and recognition ranges, including atmospheric attenuation losses. The user enters the following thermal imager parameters at the time of execution (see Figure 5.11):

Focal number ($f/\#$)
Optical efficiency (τ_a)
Optical aperture (D, in cm)
Filter bandwidth factor (K_{fd})
Scanning loss factor (K_n)
Sensor azimuth field of view (HFOV, in deg)
Sensor elevation field of view (VFOV, in deg)
Number of detectors
Frame time (T_f, in s)
Horizontal detector field of view (in mr)
Vertical detector field of view (in mr)
Observer integration time (T_e, in s)
Frame rate (F, number/s)
Integral of spectral emittance
Peak detectivity (D^*, in cm $\mathrm{Hz}^{1/2}$/W)
Limiting spatial frequency (F_{sl} in c/mr)
Critical target dimension (in m)
Target ΔT (in °C)
Atmospheric attenuation loss (in dB/km)

The overall modulation transfer function *versus* normalized spatial frequency curve (Figure 5.6) is included in the program. The program computes the MRT values corresponding to a sequence of spatial frequencies, as given in Figure 5.12.

Two different versions of the computer program were run for the data of Figure 5.11. One version computes the MRT values based on Eq. (5.45). This version resulted in the MRT values plotted as the 50 percent probability curve of Figure 5.13. The second version is exactly the same as the first, but the MRT values are divided by the square root of 14. This value is obtained using a four-bar chart display that assumes a bar length equal to seven times its width and a factor of 2 for the eye-matched filter action in the horizontal direction. The resulting MRT curve of this version is shown in Figure 5.13 as the performance model.

For comparison purposes, the NVEOL model, described in Reference [7], was also run, and the resulting curve, labeled *NVEOL*, is shown in Figure 5.13. In addition, this figure shows a set of measured data for the described thermal imaging system. From Figure 5.13 we see that the performance model agrees well with the NVEOL model, whereas the reported data follows the 50 percent probability curve somewhat better.

The program computes detection and recognition ranges, with no atmospheric loss, by multiplying the target dimension by the spatial frequency that corresponds to the target ΔT in the detection case, or by the spatial frequency divided by 4 in the recognition case. Detection and

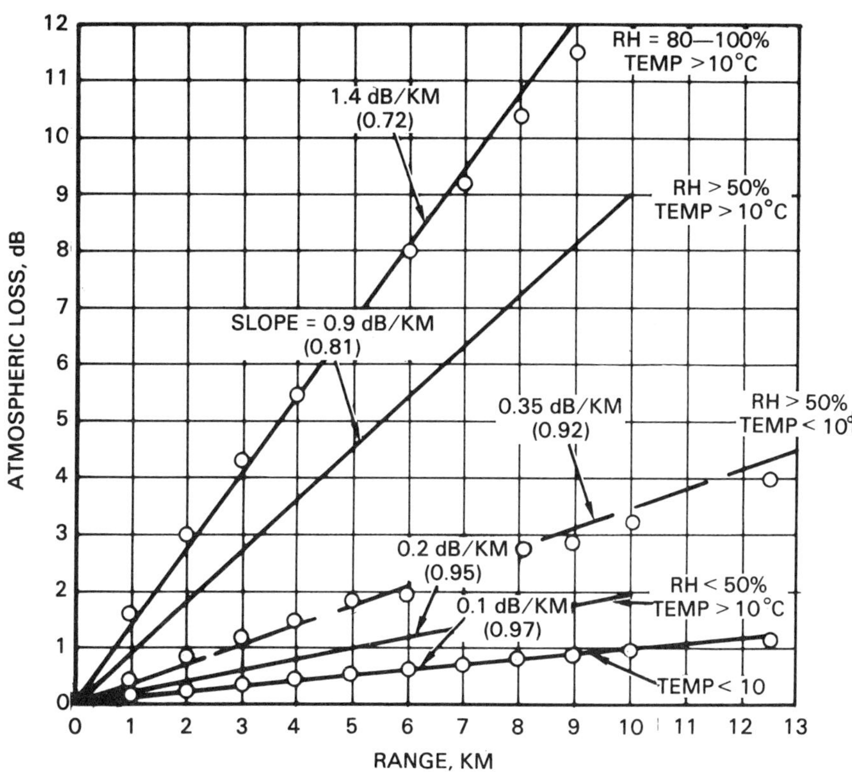

Figure 5.9 Atmospheric attenuation for clear weather with visibility range > 4 km.

recognition ranges with atmospheric loss values are computed by generating a table of values similar to those of Tables 5.4 and 5.5 and obtaining the corresponding values of detection and recognition ranges.

5.10 INFRARED SEARCH AND TRACKING SYSTEMS

The infrared search and tracking performance equation, as given by (5.13) and used in the subsequent example, was designed for a point target against a clear (nonradiating) background. The S/N ratio obtained from this equation can then be used to calculate the probability of detection for a given threshold setting by using Figure 5.2. In this section we will

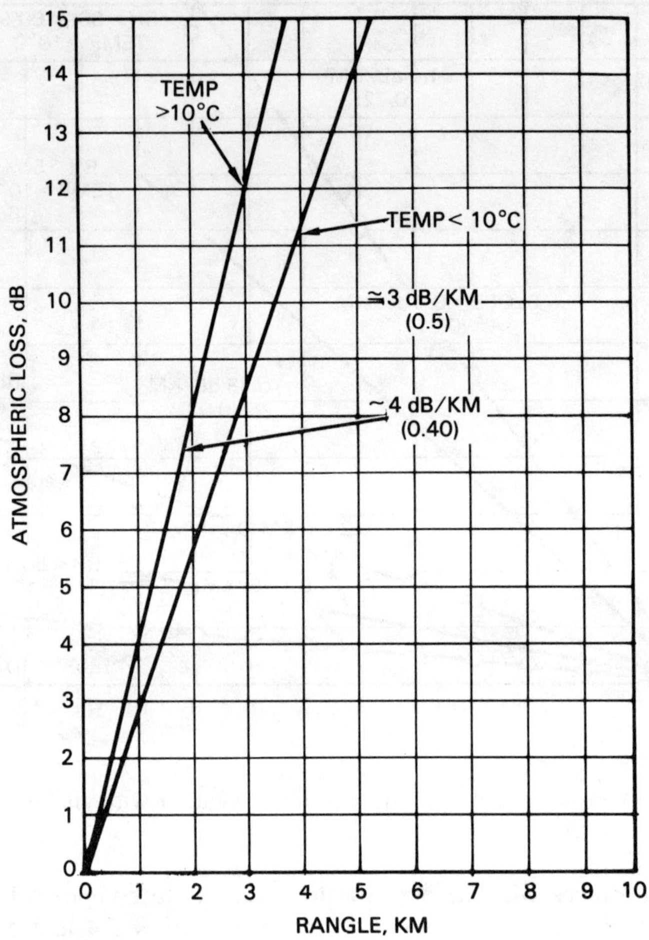

Figure 5.10 Atmospheric attenuation for fog with visibility range $\simeq$ 1 km.

extend the discussion of threshold setting and false alarm rates because this is an important consideration in IRST systems.

In a number of cases IRST systems may be used to detect targets against nonclear (radiating) backgrounds. In these cases, we can convert

		THERMAL IMAGER PARAMETERS:		
F/#	OPT EFF	APER	Kfd	Kn
1.4	0.65	1.22	1	1
OMG	NOD	Tf	ALPHA	INT TM
40 x 20	60	0.0334	2.23 x 2.98	0.2
FDOT	INTGRL	DSTAR	Fsl	TGT DIM
30	0.000084	5E + 10	0.45	2.3

Figure 5.11 Computer program showing input data.

FS	MRT
0.0045	0.00021
0.0495	0.0026
0.0944	0.00589
0.1395	0.0118
0.1845	0.02301
0.2295	0.04183
0.2745	0.10007
0.3195	0.30283
0.3645	0.86372
0.4095	1.94071

Figure 5.12 Minimum resolvable temperature values as a function of spatial frequency.

the spectral background emittance of the target and again use the S/N equation for performance calculations, even in the presence of background solar reflection and ground radiation emission.

5.10.1 False Alarm Rate and Threshold Setting

For the purposes of calculations, a relationship can be obtained between the *false alarm rate* (FAR), threshold setting, and the probability of false alarm, P_n, starting with the equation

$$P_n = \exp\left(-\frac{E_t^2}{2}\right) \tag{5.55}$$

where $E_t^2/2$ is the threshold-to-noise power ratio. The threshold-to-noise power ratio is also equivalent to the required S/N for 50 percent probability of detection:

$$P_d(50\%) \approx \frac{E_t^2}{2} \simeq S/N \tag{5.56}$$

where $P_d(50\%)$ is the 50 percent probability of detection.

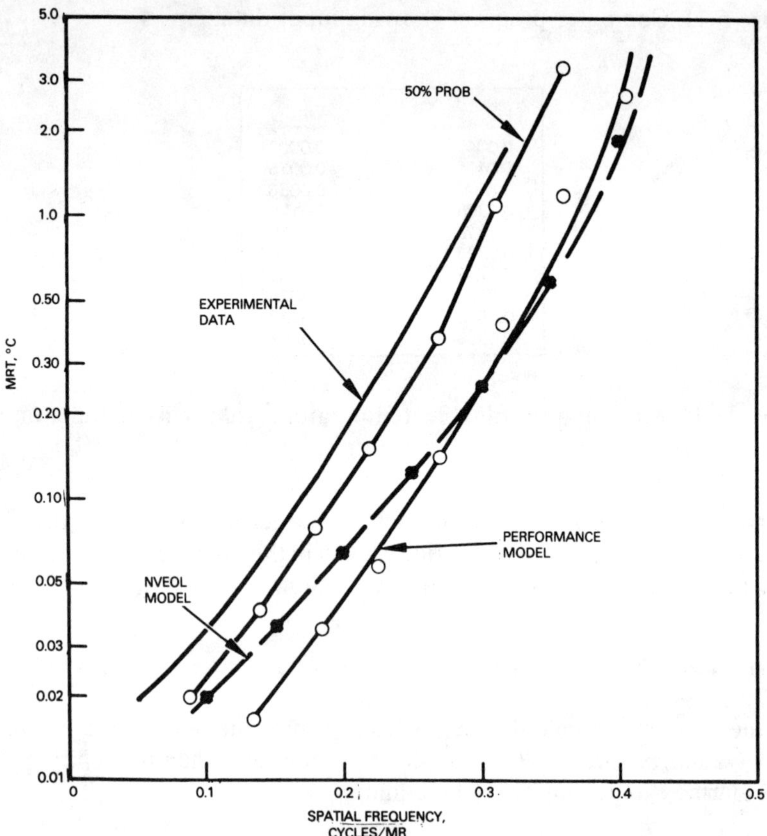

Figure 5.13 Computed MRT values of the example problem.

Substituting the value of P_n from Eq. (5.55) in (5.18), we get

$$\text{FAR} = \exp\left(-\frac{E_t^2}{2}\right)\frac{\Omega_s K_n}{\alpha_d T_f} \tag{5.57}$$

Solving Eq. (5.57) for the threshold-to-noise power ratio, we get

$$\frac{E_t^2}{2} = -\ln\left(\frac{\alpha_d T_f \mathrm{FAR}}{\Omega_s K_n}\right) \tag{5.58}$$

For a given false alarm rate, and other parameters, Eq. (5.58) will give the threshold-to-noise power ratio. Note that E_t represents the peak signal-to-rms noise voltage value and $E_t^2/2$ will represent rms signal power to rms noise power value.

5.10.2 Radiation Intensity of a Target with Background Emittance

In IRST system S/N calculations and performance analysis described in section 5.1 a non-radiating (clear) background was assumed. In a number of IRST applications, targets may be located in a radiating background. Such is the case of a tank or a truck located on a grass covered field. In these cases we can compute the effective target radiation intensity with this background, and subsequently use Eq. (5.13) for performance calculations.

The incremental target emittance was obtained using Eqs. (5.25) and (5.26). Assuming once again that the background emissivity remains constant ($\Delta\epsilon = 0$), then the spectral emittance will change only with temperature as follows:

$$\Delta W_t(\lambda) = \frac{\partial W_t}{\partial T}\Delta T \tag{5.59}$$

Equation (5.59) together with Figure 5.14 can be used to obtain the radiation intensity of a target with a given target-background temperature difference.

Consider a tank-type target with a temperature differential of 5° in the grass-covered field of Figure 5.14. For the spectral region of 8–12 μm, (the dashed area of Figure 5.14), we have

$$\int_8^{12} \frac{\partial W_t}{\partial T}\, d\lambda = 9.6 \times 10^{-5}\ (\mathrm{W/cm^2\ sr\ K}) \tag{5.60}$$

For a 5° temperature differential and a tank area of 2.5 × 6 m we can calculate the radiated power as

$$W_t = (9.6 \times 10^{-5})(5)(2.5 \times 6)(100)^2 = 72\ \mathrm{W} \tag{5.61}$$

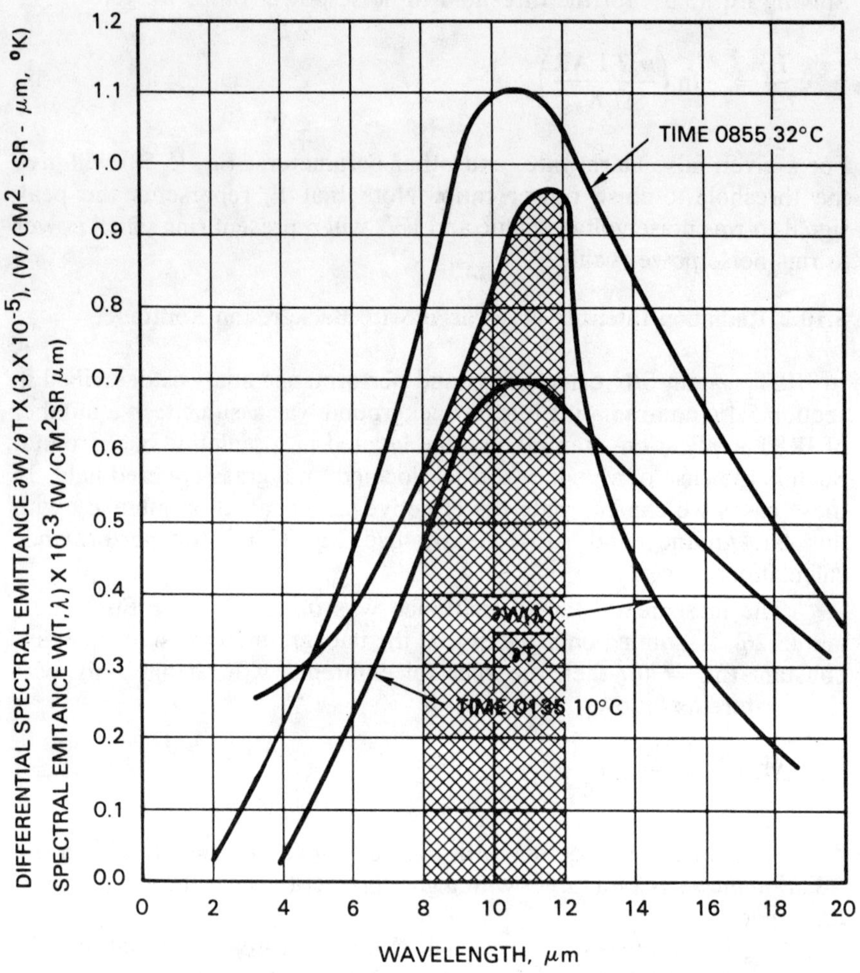

Figure 5.14 Day and night spectral emittance of grass covered field (from [5], pages 3–48).

This value of W_t corresponds to the radiation intensity J_t used in Eq. (5.13). Implicit in using J_t in Eq. (5.13) is the fact that the target is subtended by the IFOV of the detector; that is, the target can be considered a point source. The equivalent radiation intensity of the tank-type target given in Eq. (5.61) can be used in the S/N ratio equation for IRST performance analysis as described in the following example.

5.10.3 Numerical Example

Consider an IRST with the following parameters:

operating band	= 8–12 μm
target radiation intensity	= 70 W
detector field of view	= 1 mr × 1 mr
sensor field of view	= 60° (Az) × 5° (El)
sensor frame time	= 2 s
number of detectors	= 60
optics diameter	= 10 cm
focal numbers (*f*/#)	= 2
detectivity	= 5×10^{10} cm Hz$^{1/2}$/W
false alarm rate	= 1/h

$K_o = 0.65$; $K_c = 0.9$; $K_d = 0.9$; $K_m = 0.8$; $K_n = 0.8$; $K_{fd} = 1.2$

Using these parameters, obtain the 50 percent probability of detection range and the corresponding threshold setting. Also obtain the 90 percent probability of detection range.

Using the parameters, we first compute some of the parameters of the S/N Eq. (5.13) as follows:

$$L = K_o K_e K_d K_m (K_n/K_{fd})^{1/2}$$

$$= (0.65)(0.9)(0.9)(0.8)(0.8/1.2)^{1/2}$$

$$= 0.344$$

$$\Omega_s = \frac{60 \times 5}{(57.3)^2} = 0.0913 \text{ sr}$$

$$\text{S/N} = (\text{E}_t^2/2) \tag{5.62}$$

$$= -\ln \left\{ \frac{\alpha_d T_f \text{FAR}}{\Omega_s K_n} \right\}$$

$$= -\ln \left\{ \frac{(1)^2(10^{-3})^2(2)(1/3600)}{(0.0914)(0.8)} \right\}$$

$$= 18.69$$

$$= 12.72 \text{ dB}$$

Thus, the S/N for 50 percent probability of detection, or equivalently the threshold-to-noise power ratio, for the given false alarm rate and the other parameters of the IRST system will be 18.69 or 12.72 dB.

Solving Eq. (5.13) for R^2 we get

$$R^2 = \frac{\pi D J_t D^* L}{4(f/\#)(\mathrm{S/N})}\left(\frac{nT_f}{\Omega_s}\right)^{1/2} \tag{5.63}$$

Substituting the parameters of Eqs. (5.61) and (5.62) in (5.63), we get

$$\begin{aligned} R^2 &= \frac{\pi(10)(70)(5 \times 10^{10})(0.344)}{4(2)(18.69)}\left(\frac{60 \times 2}{0.0814}\right)^{1/2} \\ &= 9.16\mathrm{E}12 \quad (\mathrm{R\ in\ cm}) \\ &= 916 \quad (\mathrm{R\ in\ km}) \end{aligned} \tag{5.64}$$

From this value, the 50 percent probability of detection range will be

$$R_{50} = 30.26 \text{ km} \tag{5.65}$$

To compute the range for 90 percent probability of detection we go back to Figure 5.2. To utilize this figure, using Eq. (5.62) and the threshold setting value of 18.69. In Eq. (5.55), we compute the probability of false alarm, P_n:

$$\begin{aligned} P_n &= \exp\{-18.69\} \\ &= 7.63 \times 10^{-9} \\ &\simeq 10^{-8} \end{aligned} \tag{5.66}$$

From Figure 5.2, the 90 percent probability of detection S/N corresponding to a false alarm rate of $P_n = 10^{-8}$ is about 14.2 dB or 26.3. Since the square of the detection range is inversely proportional to S/N ratio, using Eq. (5.65), we get

$$\begin{aligned} R_{90} &= (30.26)(18.69/26.3)^{1/2} \\ &= 25.50 \text{ km} \end{aligned} \tag{5.67}$$

Thus, the 90 percent probability of detection range will be about 5 km less than the 50 percent detection probability range.

5.10.4 Atmospheric Attenuation

Atmospheric attenuation is usually expressed as a loss per km. This loss is a function of temperature, visibility, and humidity as well as spectral band of the IRST system. Given the atmospheric attenuation loss per kilometer (in dB/km), one can proceed to make a table of values as given in Table 5.6.

Table 5.6 Atmospheric Attenuation Calculations

Range with Loss (km)	*S/N Loss (dB)*	*S/N*	*SQRT of S/N*	*No Loss Range (km)*
1.0	1.0	1.258	1.122	1.122
1.1	1.1	1.288	1.135	1.248
1.2	1.2	1.318	1.148	1.377
⋮				
9.8	9.8	9.549	3.090	30.28
⋮				
12	12	15.848	3.981	47.772

The values of Table 5.6 are computed for an atmospheric loss of 1.0 dB/km one way. Multiplying this number by IRST-target range (with loss) will give the total S/N loss as given in the second column of the table. The third column of the table is the equivalent S/N. Since the IRST range is proportional to the square root of the S/N, the fourth column of the table represents the range multiplier. The product of the first and the fourth columns gives the no-loss detection range.

Thus, having obtained the no-loss detection range as given by Eq. (5.65), we can obtain the detection range with atmospheric attenuation loss from a computation as depicted in Table 5.6. For the previously computed R_{50} of 30.2 km, Table 5.6 gives a detection range with loss of 9.8 km.

Note that, in a number of cases, the overall losses through the atmosphere are known, either through calculations or by measurements. In these cases, we can reduce the S/N by this amount and calculate the corresponding detection range.

5.10.5 Mechanization Procedure in IRST Systems

Infrared search systems operate much the same way as electro-optical thermal imagers. These systems consist of an optical telescope, scanning mechanisms, detectors, and signal processing equipment. The main difference between electro-optical thermal imagers and infrared search and tracking systems lies in the use of electro-optical imagers to map ground areas or locate targets in the presence of ground reflections. Infrared search and tracking systems are usually used to locate targets in the sky or against clear backgrounds, in ground-to-air and air-to-air applications. For example, infrared systems are used in a number of missile seekers to lock-on and track targets for missile guidance and target interception.

The electro-optical imaging and the infrared systems are passive devices that transmit no energy, which prevents enemy detection and advance warning. The radiation of active microwave radars, on the other hand, can be detected by the enemy, providing a valuable tactical advantage. The passive nature of EO imaging (also referred to as FLIR) and IRST systems together with the current emphasis on stealth technology (whereby radar detection ranges are reduced by the smaller RCS of the targets) has greatly expanded the role of EO imaging and IRST systems [8]. Since infrared systems are passive devices containing only a receiver, they usually weigh considerably less than active microwave systems with both a transmitter and a receiver.

As discussed previously, the infrared systems use the temperature gradient of targets against sky or a clear background for detection. The detection ranges of these systems are considerably longer in tail attacks where target exhaust plume is seen by the sensor. The infrared wavelengths utilized in missile seekers are influenced by target radiation characteristics, atmospheric attenuation, and detector responsivity. Target radiation originates from hot metal parts of grey body radiation or combustion products of the engine. The combination of these effects reduces infrared design to five specific bands:

- 1.5 to 1.8 μm
- 2.0 to 2.5 μm
- 3.5 to 4.2 μm
- 4.5 to 4.8 μm
- 8.5 to 12.5 μm

Lead sulfide and lead selenide detectors are responsive in the 1.5–3.0 and 3.0–5.0 μm bands, respectively. These detectors were used in early IR homing missiles. Development of cooling methods, such as cryogenic coolers that reduce detector temperatures to 70 or 80 K, allowed the utilization of mercury-cadmium-telluride detectors operating in the 8–12 μm region of IR wavelengths.

Infrared target trackers use a variety of methods, with an interesting application being utilization of reticles [9]. Consider the optical system of Figure 5.15 showing target rays from a boresight target, as well as off-boresight targets. These rays are reflected from the primary mirror into a secondary mirror, converging at the detector. Placement of a "wagon-wheel" reticle (Figure 5.16) in front of the secondary mirror will modulate the incoming signal.

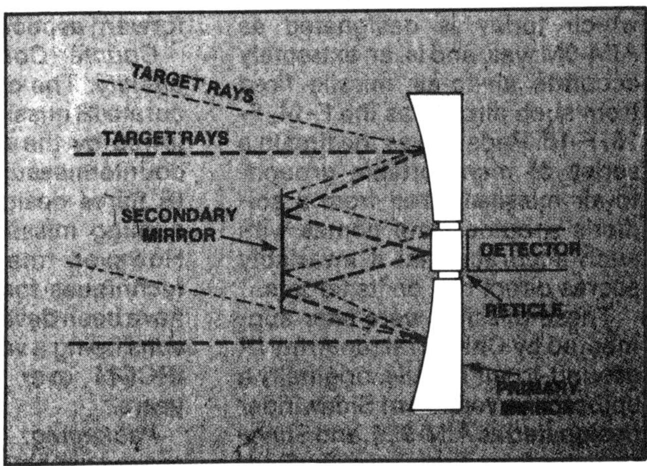

Figure 5.15 Optical ray traces from boresight and off-boresight targets.

For the boresight targets of Figure 5.16, the spinning wheel will produce equal durations of "ones and zeros" as opaque and transparent portions of the wheel intercept the target energy in equal durations. Off-boresight target of Figure 5.16 will produce unequal durations of target energy impinging on the detector, because unequal lengths of opaque and transparent portions of the wheel intercept the incoming energy.

The modulation of target return in Figure 5.16 is used through proper processing to boresight the IR set with the target and subsequent target tracking.

5.11 DETECTOR CHARACTERISTICS

Detectors used in the infrared imaging systems are generally of the photon or quantum types. These detectors operate by quantum mechanical

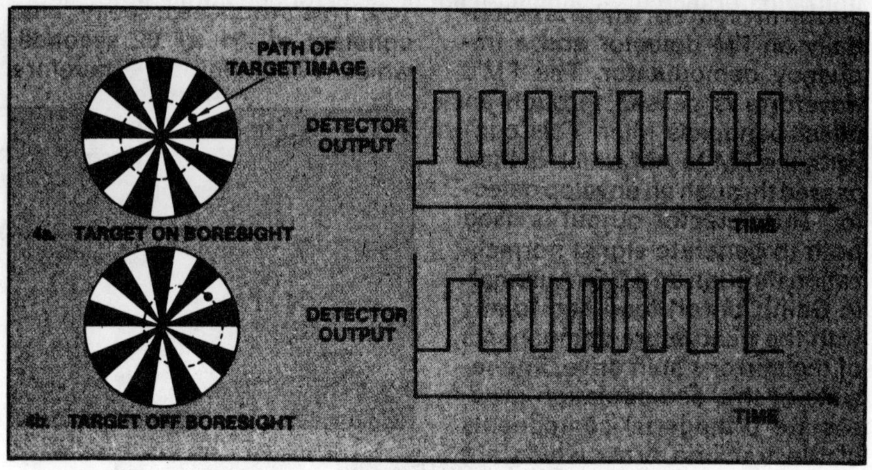

Figure 5.16 Wagonwheel reticle and target returns.

interaction between incident photons and electrons and the detector material. The detectivity of several commonly used electro-optical detectors is shown in Figure 5.17. Note that the detectivity values of the figure require cooling to about 70 K, which is achieved using Joule-Thompson cryogenic coolers.

5.12 THEMATIC IMAGING

Thematic imaging, which includes both the optical and infrared frequency spectrum, is used by ground-based, airborne, and spaceborne systems for target detection and earth surface mapping. In these imaging systems, since a variety of detectors is used, each responding to a given band of wavelengths, a great deal of information with respect to land use and ocean conditions can be collected.

Table 5.7 lists the spectral bands of the Landsat thematic mapper with the corresponding wavelengths and dominant characteristics of each band. Each band can be used individually or in combination to extract information from thematic mappers. Bands 1, 2, and 3 of Table 5.7 correspond to blue, green, and red portions of the optical spectrum. Bands 4 and higher correspond to the infrared region of the spectrum. Band 5 is used for snow and cloud differentiation, while band 6 primarily responds to the thermal properties of imaged scenes.

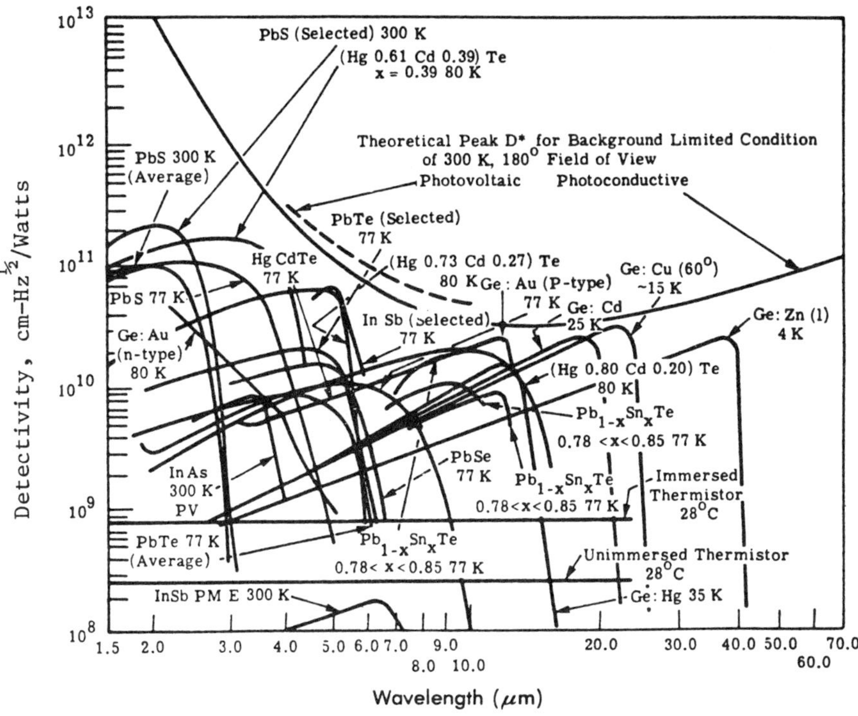

Figure 5.17 Detectivity of commonly used infrared detectors (from [2]).

Table 5.7 Thematic Mapper Spectral Band Characteristics

Band	*Range (μm)*	*Characteristic*
1	0.45–0.52	Coastal water mapping
2	0.52–0.60	Healthy vegetation
3	0.63–0.69	Chlorophyll absorption
4	0.76–0.90	Water delineation, vegetation vigor
5	1.55–1.75	Snow or cloud differentiation
6	10.4–12.5	Plant heat stress measurement
7	2.08–2.35	Hydrothermal mapping

Figure 5.18 shows a thematic image of northern Germany obtained from an airborne imager. The two scenes of the figure represent the same area of approximately 100 × 50 km. The resolution of the image is 30 ×

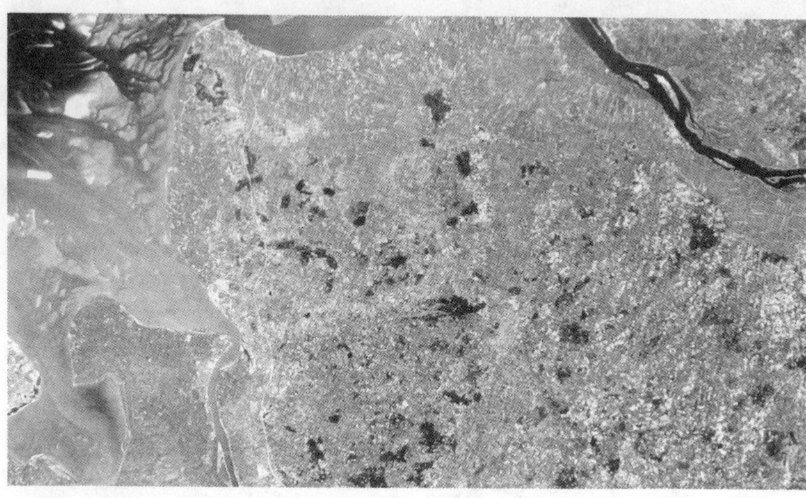

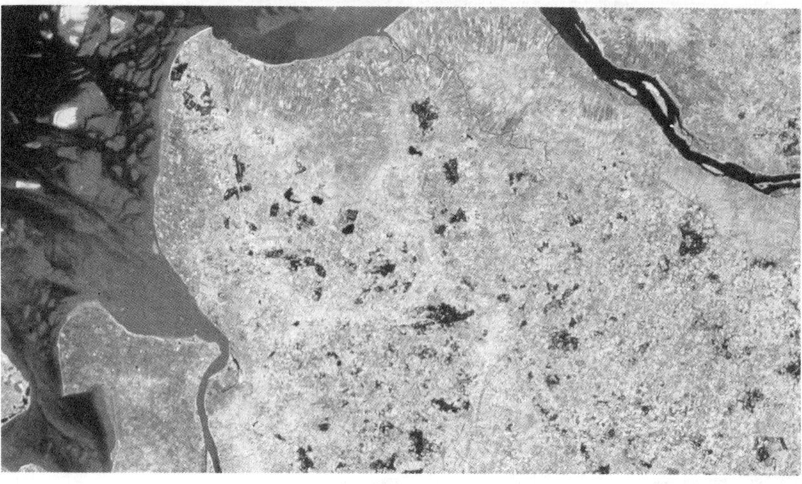

Figure 5.18 Thematic mapper image of northern Germany using data of bands 1, 2, and 3; and 3, 4, and 5. (courtesy DFVLR, Federal Republic of Germany).

30 m. In making these images a combination of bands 1, 2, and 3; and 3, 4, and 5, were used.

The upper portion of Figure 5.18 contains the response of band 1, which is used for coastal water mapping. This band can penetrate coastal waters and reveal features beneath the water surface, as shown in the figure.

Detailed studies of thematic mapper images, as exemplified by Figure 5.18, will reveal the existence and types of forests, agricultural lands, settled areas, and so on. In a recent publication by Kirchof, Mauser, and Stibig and issued by DFVLR, Federal Republic of Germany, excellent recognizability of land use was reported using bands 2, 4, and 5 of Table 5.7.

5.13 SUMMARY

In this chapter, the system-oriented design and performance equations of electro-optical thermal imaging systems were derived and applied to numerical examples. New methods of including atmospheric attenuation characteristics in performance equations were developed and demonstrated. In writing this chapter, the best features of the corresponding sections of the standard electro-optical reference books of Seyrafi [3] and Lloyd [4], as well as the computer program of Ratches [7], were used. The analysis was carried beyond the contents of these references in streamlining solutions for digital computers.

REFERENCES

1. S. A. Hovanessian, *Radar System Design and Analysis*, Artech House, Dedham, MA, 1984, Chapter 6.
2. W. L. Wolfe and G. J. Zissis, *The Infrared Handbook*, Environmental Research Institute of Michigan, Ann Arbor, 1978.
3. K. Seyrafi, *Electro-Optical System Analysis*, Electro-Optical Research Corporation, Los Angeles, CA, 1981, Chapter 7.
4. J. M. Lloyd, *Thermal Imaging System*, Plenum Press, New York, 1975, Chapter 5.
5. J. Johnson, *Image Intensifier Symposium*, Paper AD-220160, Fort Belvoir, VA, 1958.
6. L. P. Obert et al., *Visionics EO Sensor Performance*, DCS Corporation, Arlington, VA, June 1981.
7. J. A. Ratches et al., *Night Vision Laboratory Static Performance Model for Thermal Viewing Systems*, R and D Technical Report, ECOM-7043, April 1975.
8. "Emphasis on Stealth Will Spur Improvement in Flir Sensors," *Aviation Week and Space Techn.*, September 7, 1987.
9. J. J. May and M. E. VanZee, "Electro-Optic and Infrared Sensors," *Microwave J.*, September 1983, pp. 121–131.

Chapter 6
Laser Radar Systems

This chapter contains a general discussion of the design and performance equations of laser radar systems. The signal power and noise power equations are derived for both direct and heterodyne laser radar systems. Probability of detection curves are given for glint and speckled targets, and methods of calculating the number of false alarms are discussed.

Ground backscattering and natural interference signals—such as emittance from sky, clouds, and ground—are discussed, and methods of incorporating these signals in detection calculations are outlined. Numerical examples are given to illustrate the application of the given equations to the design of laser radar systems.

Laser radars, because of their narrow beamwidths, usually have poor search capability. These systems, however, are used in conjunction with passive infrared search and tracking systems IRST and thermal imaging systems to obtain target range and reduce false alarms (detections). In these applications IRST or thermal imagers locate the potential targets in large volume search. Subsequently, the laser radar system is pointed in the direction of the target to obtain range. In this application the laser system can also confirm the existence of the target. Laser radars can also be used in conjunction with microwave and MMW radars in a similar manner to improve angular position and range accuracy of the targets.

A number of equations and figures used for performance analysis of microwave radar systems also apply in the case of laser radar systems. For continuity of presentation and the convenience of the reader, these items are repeated in this chapter

6.1 INTRODUCTION

Laser radar systems are active devices that operate in a manner very similar to microwave radars but at a much higher frequency. This higher frequency allows using smaller components and has remarkable angular accuracy. The atmospheric attenuation losses, however, are considerable at these high frequencies.

Ground-based laser systems usually have a limited detection range (about 10 km), primarily because of atmospheric effects. Spaceborne laser radar systems, however, can be built with detection ranges of thousands of km in the absence of atmospheric attenuation.

Laser radars measure radar target range and angular positions in much the same way as microwave radars. The small angular beam spread of laser radars (in μrad) limits the performance of these radars in large volume searches. In a number of cases, large volume search devices, such as microwave radars and infrared systems, are used as initial search and target locating devices. This coarse target angular position is subsequently used for cuing the laser radar systems.

Primarily because of atmospheric attenuation windows and the availability of components, laser radars are built at specific wavelengths. Figure 6.1(a) shows atmospheric windows at laser radar wavelengths. From this figure we see transmission coefficients of unity at the 1 and 10 μm regions. Most of the operating laser radars are constructed at 1.06 and 10.6 μm wavelengths. The 1.06 μm lasers are referred to as Neodymium YAG crystal lasers, while 10.6 μm laser radars are called CO_2 gas lasers. The solid state laser system of 1.06 μm is not as efficient as the CO_2 gas lasers of 10.6 μm. The former may have an efficiency of 1 or 2 percent, while the latter may have an efficiency of about 10 percent.

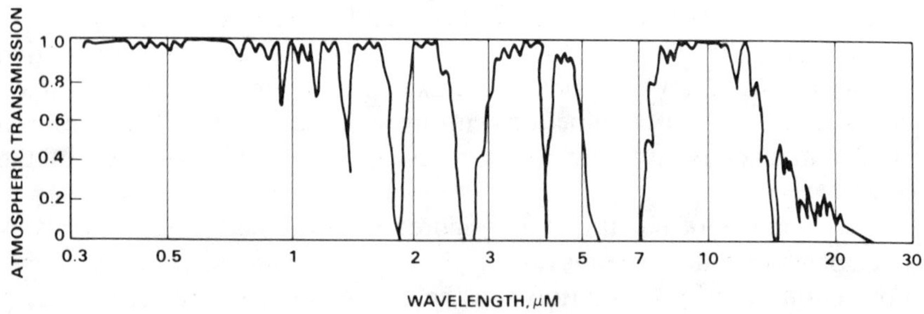

Figure 6.1(a) Electromagnetic radiation through the atmosphere at micron wavelengths.

6.2 PERFORMANCE CALCULATIONS

Referring to the geometry of Figure 6.1, the signal power received by a laser radar from a target at range R can be written as in Eq. (1.34):

$$S = \underset{①}{\frac{P}{4\pi R^2}} \cdot \underset{②}{\frac{4\pi}{(\pi/4)\theta_{BW}^2}} \cdot \underset{③}{\frac{\pi}{4}(R\theta_{BW})^2} \cdot \underset{④}{\rho} \cdot \underset{⑤}{\frac{1}{2\pi R^2}} \cdot \underset{⑥}{A} \cdot \underset{⑦}{\tau_o} \tag{6.1}$$

where

S = target signal power at the laser radar
P = transmitted pulsed power
R = radar-target range
θ_{BW} = laser beamwidth
ρ = target backscattering coefficient
A = lens aperture area
τ_o = optical efficiency

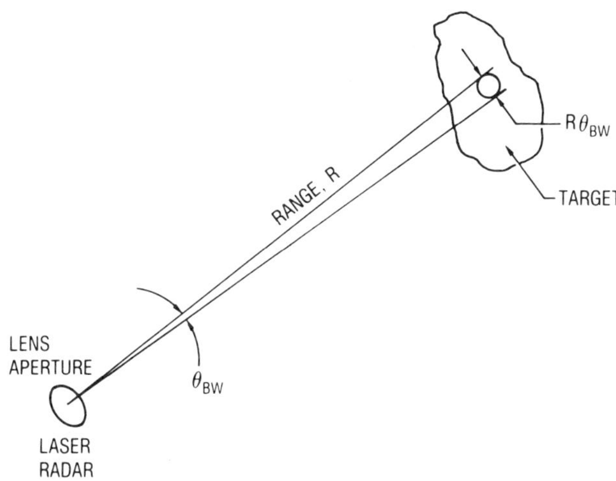

Figure 6.1 Laser radar target geometry.

The first term of Eq. (6.1) represents the distribution of the point source power P over a spherical area of $4\pi R^2$. This would be the power distribution of a point source without the benefit of the lens that focuses the power in a given direction. The second term is a measure of focusing effect of the lens where it is used to direct the radiated power in a given direction. This gain is the ratio of the spread of power over a sphere of 4π sr and the laser beamwidth, also expressed in sr.

The third term of Eq. (6.1) represents the area of the target. In this derivation, the laser beam is assumed to be considerably smaller than the target area. This assumption can be removed if the target is smaller than the laser beam spread at radar target range, R, as will be shown later.

The fourth term is the backscattering coefficient of the target. This value depends on the surface condition of the target, such as roughness, wetness, and so on. The fifth term is the mathematical expression of the fact that the power at the target is also spread over a hemispherical area similar to the transmitted power of the first term. In Eq. (6.1), the power at the target is assumed to spread over the forward hemisphere of 2π sr. The resulting power density at laser range R will be $1/2\pi R^2$.

Another often-used assumption, with respect to the distribution of power at the target, is that of Lambertian scattering: Lambertian distribution assumes that the reflected flux per unit solid angle is proportional to the cosine of the angle between the direction of interest and the normal to the surface. Referring to Figure 6.2, we can compute the "sr" spread of target power as

$$\int_o^{\pi/2} 2\,\pi \sin\theta \cos\theta \, d\theta = [2\pi\,(\sin^2\theta/2]_o^{\pi/2}$$

$$= \pi \tag{6.2}$$

Thus, in the case of Lambertian assumption, the fifth term of Eq. (6.1) will be divided by π instead of 2π.

Target return power is intercepted by the lens area A as given by the sixth term of Eq. (6.1). The seventh term represents the optical efficiency of the laser radar transmission chain following the front aperture of the lens. The target signal power of Eq. (6.1) can be simplified to the forms of Eqs. (6.3a) and (6.3b) for Lambertian and semispherical target power distribution, respectively.

$$S = \frac{PA\tau_o\rho}{\pi R^2} \tag{6.3a}$$

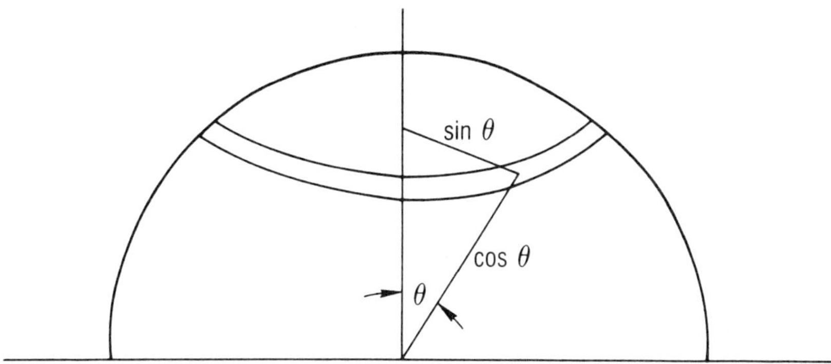

Figure 6.2 Lambertian distribution of incident power at the target.

$$S = \frac{PA\tau_o\rho}{2\pi R^2} \tag{6.3b}$$

According to Eq. (6.3), the signal power of the target is inversely proportional to the square of the laser radar target range, R. Note that in the derivation of Eq. (6.3), we assumed that the laser beam is smaller than the target. We will see in the following analysis that in cases where the laser radar beam is larger than the target (point target), the signal power is inversely proportional to the fourth power of the laser-target range. This is the same as in the case of microwave radars.

6.3 LASER BEAM LARGER THAN TARGET

If the target does not intercept the entire laser beam as expressed by the third term of Eq. (6.1), this term should be substituted by the laser *radar cross section* (RCS). This target cross section is denoted by σ; Eq. (6.1) becomes

$$S = \frac{2PA\sigma\rho\tau_o}{\pi^2 R^4 \theta_{BW}^2} \tag{6.4}$$

Equation (6.4) can be described in terms of laser radar parameters by substituting the laser radar beamwidth in terms of lens diameter and wavelength. The expression for beamwidth can be written

$$\theta_{BW} = \frac{\lambda}{D} \tag{6.5}$$

where λ is the wavelength and D is the lens diameter.

At this point, a discussion of beamwidths expressed as a function of lens dimensions is appropriate. In microwave radars, the 3-dB or one-half power beamwidth usually is applied. This value for a sin x/x beam structure is equal to 0.88 λ/D, expressed in rad. In optical systems, $1/e = 0.367$ instead of one-half power point is used. This power level will result in a beamwidth of 1.05 λ/D, again expressed in rad. From these values, note that the usual beamwidth approximation of λ/D can be considered appropriate for both microwave and laser radars.

Using Eq. (6.5) and expressing the lens area as

$$A = \frac{\pi}{4} D^2 \tag{6.6}$$

in Eq. (6.4), we get

$$S = \frac{8PA^2\sigma\rho\tau_o}{\pi^3 R^4 \lambda^2} \tag{6.7}$$

From Eq. (6.7), note that the target signal is inversely proportional to the fourth power of laser radar target range.

6.4 FLOODLIGHT LASER RADAR SYSTEMS

In a number of applications, two different laser radars are used for transmitting and receiving functions: target illumination (i.e., transmission of energy toward the target), and receipt of portions of this energy scattered from the target. Figure 6.3 shows the geometry of floodlight lasers. In these lasers, the transmitting beamwidth usually is larger than the receiving beamwidth, hence the name *floodlight lasers*. This larger transmitting beam circumvents the problem of target search.

The target signal power, Eq. (6.1), can be adapted easily to floodlight laser radars as follows:

$$S = \frac{P}{4\pi R_1^2} \cdot \frac{4\pi}{(\pi/4)\,\theta_{BW1}^2} \cdot \sigma \cdot \rho \cdot \frac{1}{2\pi R_2^2} \cdot A_2 \cdot \tau_o \tag{6.8}$$

where subscripts 1 and 2 are used for transmitting and receiving functions. In Eq. (6.8), we assumed that the transmitting beam is considerably larger than the target area at range R_1. The transmit lens beamwidth in terms of lens area can be written as

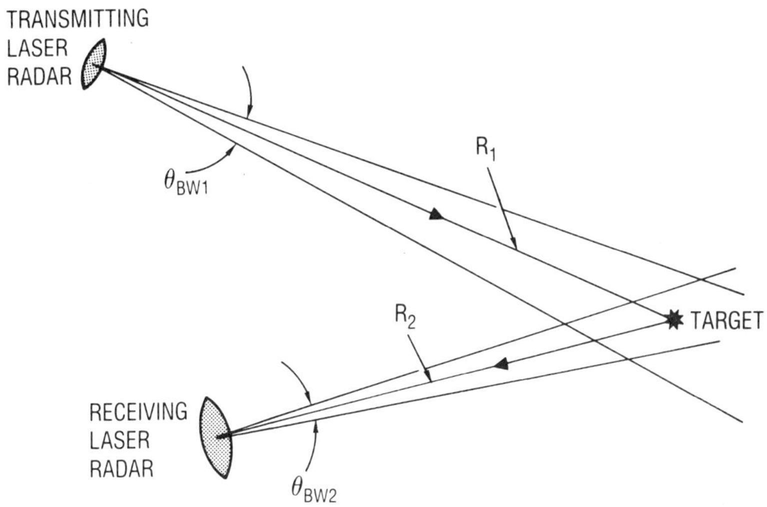

Figure 6.3 Floodlight laser radar transmit-receive systems.

$$\theta_{BW1}^2 = \frac{\lambda^2}{D_1^2} = \frac{\pi\lambda^2}{4A_1} \tag{6.9}$$

in Eq. (6.8); simplifying, we get

$$S = \frac{8PA_1A_2\sigma\rho\tau_o}{\pi^3\lambda^2R_1^2R_2^2} \tag{6.10}$$

This signal power equation is similar to the *semiactive* microwave radar performance equation. Note the presence of transmitting and receiving apertures and the corresponding laser radar target ranges in Eq. (6.10). Also note that the laser radar cross section of the target σ in Eq. (6.10) is the cross section seen by the receiving laser radar. This cross section usually has a different value than the target cross section where both the transmitter and the receiver are at the same location.

6.5 LASER RADAR RECEIVERS

Laser radar receivers are of two types: direct and heterodyne detection [Refs. 1, 2]. Direct receivers are similar to electro-optical detectors where the backscattered energy from the target impinges on the detector and is converted to an electrical voltage or an electrical current. Heterodyne receivers, on the other hand, contain a stable local oscillator which is used for frequency reference between transmit and receive waveforms.

In heterodyne receivers, the received waveform from the target is heterodyned (multiplied) by the transmitted frequency to reduce the frequence of the returned waveform. At these reduced frequencies, the target signal can be amplified, enhancing the detection process. Note that in heterodyne laser radar, the coherency (phase relationship) of transmitted and received signals is preserved. This allows calculation of the laser radar-target closing rate $\mathring{R}$ by doppler processing. Direct detection lasers operate with noncoherent signals where $\mathring{R}$ calculations are not possible.

Figure 6.3(a) shows the block diagram of direct and heterodyne laser radar receivers. In the heterodyne receiver, the transmitted waveform is generated by a continuous wave (CW) source. This CW waveform is used as a reference to reduce the frequency of the received waveform from the target.

The detector noise level for *direct* detection is given as

$$N = (A_d \Delta f)^{1/2}/D^* \tag{6.10(a)}$$

where

N = noise level, W
A_d = detector area, cm^2
Δf = receiver bandwidth, Hz
D^* = detectivity, cm − Hz$^{1/2}$/W

The detector area, as seen from Figure 6.4 for a square detector, is related to this lens diameter and focal length as follows:

$$\begin{aligned} A_d &= \alpha_d^{1/2} f_1 \cdot \alpha_d^{1/2} f_1 \\ &= \alpha_d f_1^2 \end{aligned} \tag{6.11}$$

where A_d is the detector area, in cm^2. The focal length f_1 is given

$$f_1 = (f/\#)D \tag{6.12}$$

where $f/\#$ is the lens focal number. The value of $\alpha_d^{1/2}$ of Eq. (6.12) is the instantaneous detector field of view (IFOV), and A_d is the detector area.

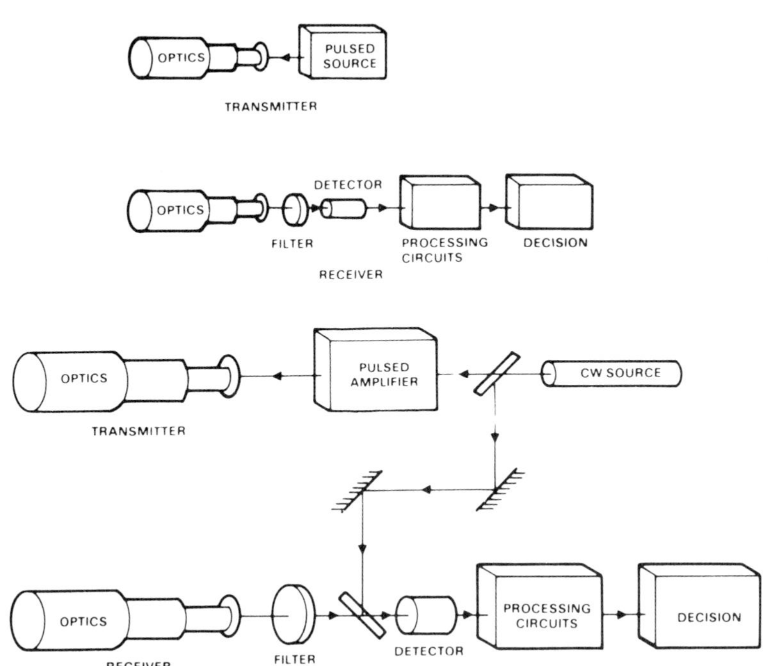

Figure 6.3(a) Laser receiver systems.

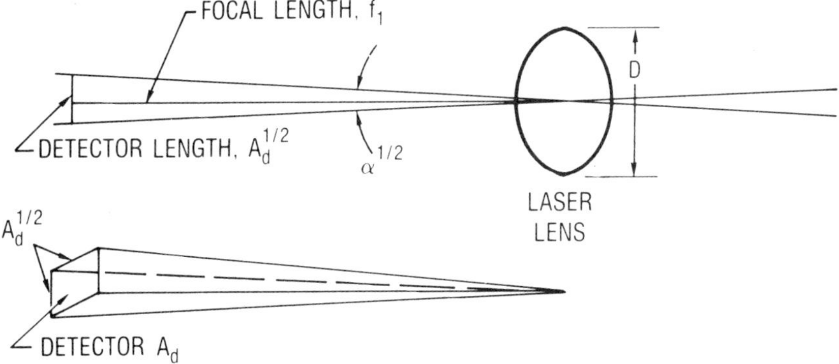

Figure 6.4 Detector area and location with respect to laser radar lens.

The noise level of a *heterodyne* receiver, as given in Reference [3], can be computed as follows. Quantum mechanical analysis has shown that an ideal amplifier has a noise power spectral density of

$$\psi(f) = \frac{hf}{\exp(hf/kT) - 1} + hf \tag{6.13}$$

where

$\psi(f)$ = spectral density, W/Hz
h = Planck's constant
= 6.6256×10^{-34} W-s²
f = frequency, Hz
k = Boltzmann constant
= 1.38×10^{-23} W-s/K
T = absolute temperature, K

Equation (6.13) can be evaluated as a function of wavelength, using the relation

$$\lambda = \frac{c}{f} \tag{6.14}$$

where c is the velocity of propagation and f is the frequency. The noise power plot of Eq. (6.13) is shown in Figure 6.5. The value of $\psi(f)$ of this figure is that of Eq. (6.13). This figure also includes curves of

$$\mu(f) = hf \tag{6.15}$$

and

$$\gamma(f) = kT \tag{6.16}$$

Note that at cm wavelengths, with frequencies corresponding to microwave radars, the noise power is given by Eq. (6.16). At μm wavelengths, the noise power is more correctly approximated by Eq. (6.15). Using this, the noise level of a *heterodyne* receiver of a laser radar system can be written as

$$N = (fh/\eta)(B) \tag{6.17}$$

where η is the quantum efficiency and B is the receiver bandwidth. Quantum efficiencies usually are on the order of 30–50 percent.

6.6 PROBABILITY OF DETECTION: GLINT TARGETS

The previous equations can be used to compute the target S/N ratio of a given laser radar against a specified target. This S/N ratio can be used to compute the probability of detection with threshold setting as a parameter. Before discussing the probability of detection, we shall describe types of target returns encountered by laser radar systems.

Laser radar targets can be divided into two general categories: speckled targets and glint targets. The returns of these targets and their statistical behavior are discussed extensively in Reference [4]. In summary, we can categorize such targets as having either several reflecting points or a single dominating reflector. Targets consisting of a single dominating reflector (such as corner reflectors or normal surfaces) usually give target returns from pulse to pulse that are more or less constant. These targets are referred to as *glint targets*. A histogram plot of glint targets will reveal a Gaussian distribution around the dominating reflector. The standard deviation of this distribution usually is small as compared

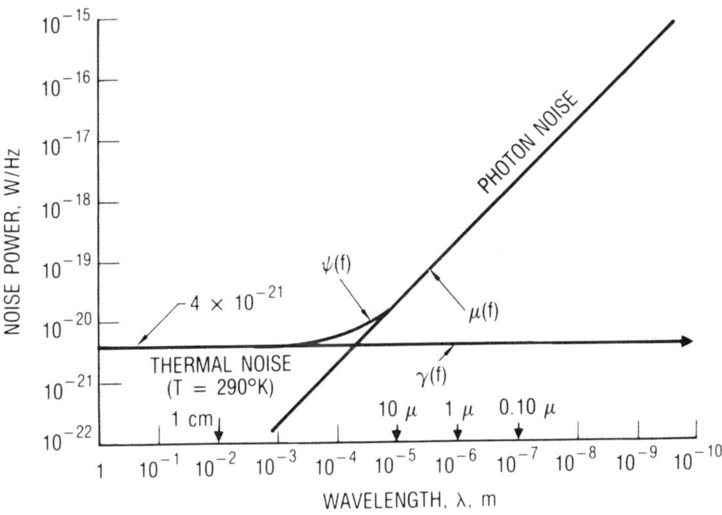

Figure 6.5 Noise level of an ideal amplifier as a function of wavelength.

to its average value. Figure 6.6 gives a normalized histogram of a glint target return. In this case, the target consisted of a retroreflector. One hundred pulse returns were used in the construction of this figure. The average value is 1 with a $\sigma = 0.067$.

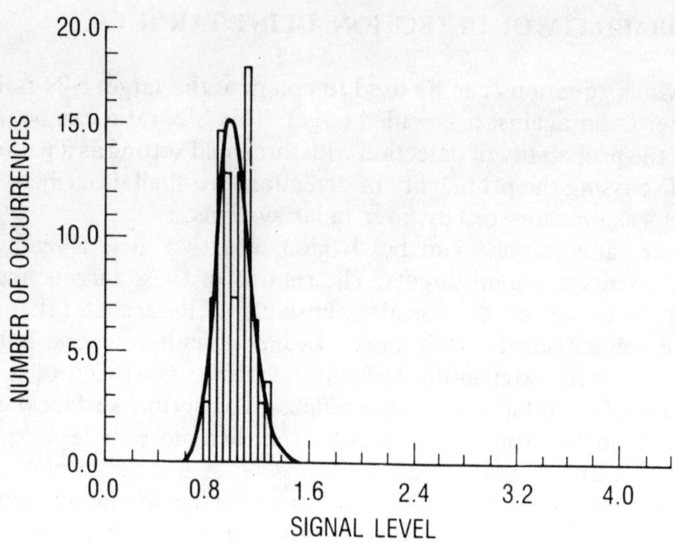

Figure 6.6 Histogram of a glint target signal power showing a Gaussian probability distribution (average = 1.0; standard deviation ≃ 0.067).

When we have the S/N ratio of a glint target, we can compute the probability of detection of this target using the curves of Figure 6.6(a). These curves are also used in microwave radars and give the probability of detection of a sinusoidal signal in Gaussian noise. The figure includes the probability of false alarm as a parameter. This parameter is used to set the receiver threshold as a function of desired false alarm rate, as will be discussed in Section 6.8.

A precise analysis of the probability of detection of glint targets requires a small correction to account for the spread of the target signal around the average value, as given in [5], page 157. For most cases, however, this correction is less than 1 dB in signal power.

6.7 PROBABILITY OF DETECTION: SPECKLED TARGETS

Targets with several reflecting points, each with somewhat equal backscattering characteristics, tend to give speckled power returns. Assume that the power return from each reflecting point behaves with a Gaussian probability distribution. The sum of the scattered target power, representing the total signal power from the target, can then be represented by a Rayleigh distribution.

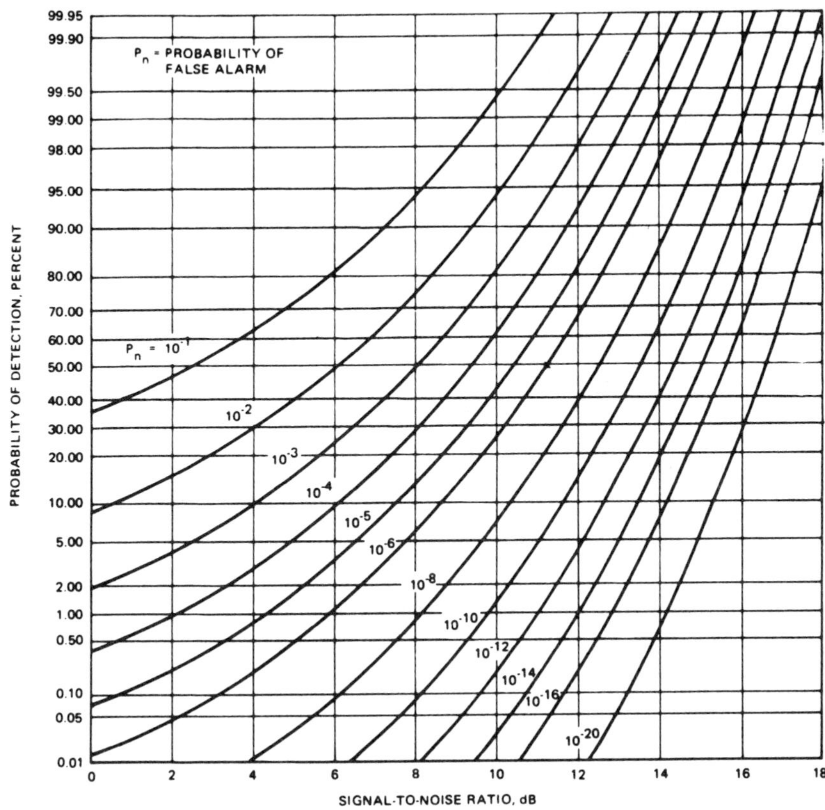

Figure 6.6(a) Probability of detection as a function of signal-to-noise ratio for glint targets.

Figure 6.7 shows a histogram consisting of 1200 pulse returns obtained from a speckled target simulated by a flame-sprayed, aluminum calibration plate [6]. Speckled targets also result in radar returns approximating exponential distribution. Both Rayleigh and exponential distribution can be used in conjunction with the probability of detection curves of Figure 6.6(a) to obtain the probability of detection of speckled targets. Reference [5], page 153, shows that a negligible difference in the probability of detection exists, whether Rayleigh or exponential distribution is used in these calculations. Figure 6.7(a) shows the probability of detection curves as a function of signal-to-noise ratio for a speckled target. The probability of false alarm is used as a parameter. Comparison of these curves with the glint target curves of Figure 6.6(a) will show

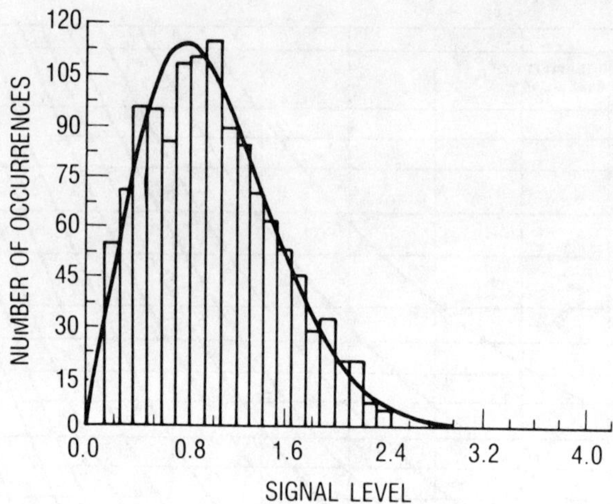

Figure 6.7 Histogram of a speckled laser target power showing a Rayleigh distribution.

that higher levels of signal-to-noise ratio are required for speckled targets to achieve above 30 percent probability of detection values. Note that the same probability of detection curves in Figures 6.6(a) and 6.7(a) are used in the case of microwave radars. In microwave radars, glint targets are referred to as nonscintillating targets, and speckled targets are referred to as scintillating targets.

6.8 PROBABILITY OF FALSE ALARM AND TIME BETWEEN FALSE ALARMS

The previously given probability of detection curves specify the probability of detection of targets as a function of S/N ratio with the probability of false alarm as a parameter. The probability of false alarm is used to set the laser radar receiver threshold as well as to calculate the number of false alarms per unit time or, equivalently, time between false alarms.

The threshold setting for a given false alarm corresponds to a 50 percent probability of detection S/N ratio of a glint target. For example, from Figure 6.6(a), the threshold setting for a probability of false alarm of 10^{-6} will correspond to 11.25 dB. This value is the effective power-to-noise ratio

$$\frac{E_t^2}{2} = 11.25 \text{ dB} \tag{6.18}$$

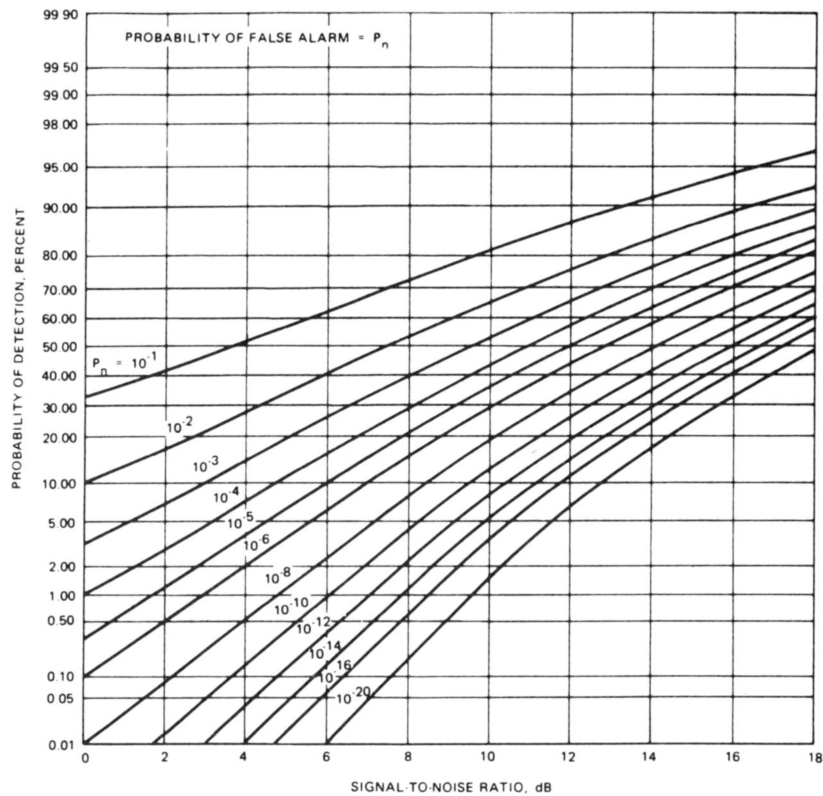

Figure 6.7(a) Probability of detection as a function of signal-to-noise ratio for speckled targets with exponential or Rayleigh distributions.

where E_t is the peak threshold-to-noise voltage ratio. From Eq. (6.18), we have

$$E_t = 5.16 \tag{6.19}$$

To compute the number of false alarms per unit time (s), consider the following discussion. Assume a pulsed, laser radar system with a pulse duration of τ s and range gates of the same duration as shown in Figure 6.8. The number of range gates, n, examined during 1 s for the presence of targets will be

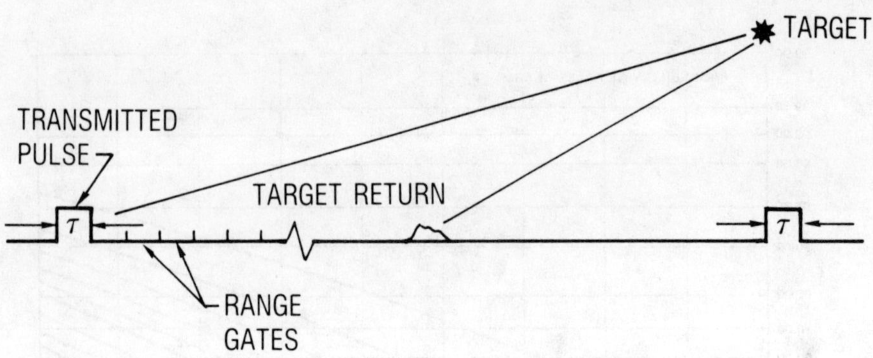

Figure 6.8 Transmitted pulses, range gates, and target returns.

$$n = \frac{1}{\tau} \tag{6.20}$$

The probability of seeing a false target in each of these range gates is equal to the probability of false alarm, P_n. Thus, the number of false alarms/s, N_{fa}, can be written

$$N_{fa} = \frac{P_n}{\tau} \tag{6.21}$$

Note that these false alarms are caused by receiver noise only. They are not produced by outside environmental effects, such as clouds, ground emissivities, or weather conditions.

Equation (6.21) gives the number of false alarms in 1 s produced in a random range gate. If we impose the additional condition that a *false alarm* will be considered a *false target detection* if, and only if, it appears in a specified range gate following two consecutive pulse transmissions, we have

$$N_{fa} = \frac{P_n^2}{\tau} \tag{6.22}$$

This additional restriction, which is easy to implement, will considerably reduce the number of false detections/s.

As an example of Eq. (6.22), consider a laser radar with a pulse duration of 1 ns and a probability of false alarm of 10^{-6}. When these values are used, Eq. (6.22) results in

$$N_{fa} = \frac{(10^{-6})^2}{10^{-9}} = 10^{-3} \tag{6.23}$$

or one *false target detection* every 16.67 min.

6.9 LASER RADAR MEASUREMENTS

Laser radars are capable of measuring radar-target range, range rate and angular position much the same way as other microwave radars. The laser radar target range, R, is calculated by measuring the elapsed time, t, between transmit and receive pulses, to yield

$$R = \frac{ct}{2} \tag{6.24}$$

Heterodyne lasers can measure radar target range rate, $\dot{R}$, by doppler frequency processing:

$$\Delta f = \frac{2\dot{R}}{\lambda} \tag{6.25}$$

where Δf is the measured doppler shift between transmit and receive signals and λ is the wavelength. The angular position of the target is obtained by registering the angular position of the laser beam at the time of target signal return. Target angular position of a laser radar can be further improved by the use of quadrant detector configuration (see Reference [7]). In this mechanization the returned laser pulse impinges on four detectors placed in a quadrant formation. This scheme is similar to monopulse radar detection discussed in Chapter 2. The output of each of four detectors is proportional to the target image power incident on it. Using the resulting detector power levels, a more accurate target angular position location can be computed.

Other range measurement techniques used in microwave radars, such as frequency modulation ranging and multiple PRF ranging, can be used equally well in laser radars (see Chapter 2). In fact, Reference [8] gives a detailed discussion of the mechanization procedure of a laser radar using pulse compression methods. In this mechanization, a CO_2 heterodyne laser radar system is used to obtain detection ranges of several kilometers with a ranging accuracy of 10 m and a radial velocity accuracy of less than 1 m/s.

6.10 NUMERICAL EXAMPLE

We wish to design a 10 km ground-based, imaging laser radar that satisfies the following requirements:

Coverage	0.5 × 0.5°
Resolution	2 × 2 ft
Range	10 km

We anticipate that this laser radar will produce images similar to the one given in Figure 6.8(a) of a tank. The design of this laser radar can be accomplished through a series of calculations and a series of parameter estimations. These parameters are then used in S/N ratio and probability of detection calculations.

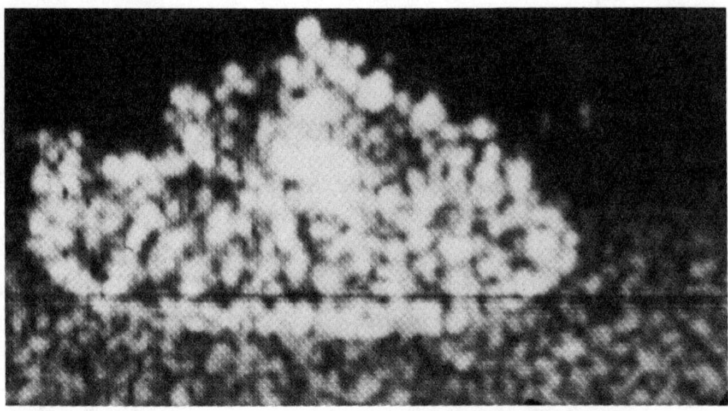

Figure 6.8(a) Anticipated laser radar image of a tank type target.

The 2-ft resolution requirement will give rise to a detector IFOV of

$$\alpha_d^{1/2} = \frac{2}{10{,}000 \times 3.28} = 6.09 \times 10^{-5} \text{ rad} \quad (6.26)$$

where 10,000 m is the range and 3.28 is the conversion from m to ft. Using Eq. (6.26):

$$\alpha_d = 3.71 \times 10^{-9}\ \text{sr} \tag{6.27}$$

The coverage of $0.5 \times 0.5°$ can be converted to sr as follows:

$$\begin{aligned} \Omega &= \frac{0.5 \times 0.5}{(57.3)^2} \\ &= 7.6 \times 10^{-5}\ \text{sr} \end{aligned} \tag{6.28}$$

Round-trip travel time for 10 km coverage becomes

$$\begin{aligned} t &= \frac{2R}{c} \\ &= 6.67 \times 10^{-5} \end{aligned} \tag{6.29}$$

The number of resolution cells or pixels N_p required for the coverage of the target can be computed from Eqs. (6.27) and (6.28) as follows:

$$\begin{aligned} N_p &= \frac{\Omega}{\alpha} \\ &= 2.048 \times 10^4 \end{aligned} \tag{6.30}$$

Assuming a search time of 0.1 s for the $0.5 \times 0.5°$ coverage, we find the number of pixels covered per second, $\dot{N}_p$, to be

$$\begin{aligned} \dot{N}_p &= \frac{N_p}{0.1} \\ &= 2.048 \times 10^5\ \text{pixels/s} \end{aligned} \tag{6.31}$$

The number of pixels for the round trip time of 10 km can be computed using Eqs. (6.29) and (6.31):

$$\begin{aligned} N' &= \dot{N}_p t \\ &= 13.66 \\ &\simeq 15 \end{aligned} \tag{6.32}$$

This number also represents the number of receivers required to cover the ten kilometer design range. As the lasar radar beam covers new search areas, target return pulses from previously covered areas will be arriving and there should be receivers to receive them. For a known target range, the number of receivers can be reduced to cover the round-trip delay of the range. The number of receivers can also be reduced by increasing the search time.

6.11 S/N RATIO COMPUTATIONS

Consider the design of a CO_2 laser radar used in section 6.10. The S/N ratios can be computed from Eqs. (6.3a) and (6.17) and written as follows:

$$\mathrm{S/N} = \frac{PA\rho\tau_o}{\pi R^2(hf/\eta)(B)} \tag{6.33}$$

where we assume that the laser radar beam covers the target, as it should in the imaging laser radar of this design.

Consider the following design parameters for the imaging laser radar:

P = pulse power
= 20 W
D = aperture diameter
= 8 in
ρ = target reflectivity
= 0.01
τ_o = optical transmission
= 0.40
h = Planck's constant
= 6.6256×10^{-34} W-s²
f = frequency
= c/λ
c = velocity of propagation
= 3×10^{10} cm/s
λ = wavelength
= 10.6 μm
η = quantum efficiency
= 0.50
B = receiver bandwidth, Hz

These design parameters can be used directly in the S/N ratio, Eq. (6.33). It will be more convenient, however, to compute some of these parameters beforehand. The aperture area, A, can be computed from lens diameter

$$\begin{aligned} A &= \frac{\pi}{4} D^2 \\ &= \frac{\pi}{4} (8 \times 2.54 \times 10^{-2})^2 \\ &= 0.0324 \text{ m}^2 \end{aligned} \tag{6.34}$$

The transmitted frequency, f, is as follows:

$$\begin{aligned} f &= c/\lambda \\ &= \frac{3 \text{ E8}}{10.6 \text{ E} - 6} \\ &= 2.83 \text{ E13 Hz} \end{aligned} \tag{6.35}$$

The receiver bandwidth, B, can be computed from a consideration of on-target-time:

$$B = \frac{1}{\text{OTT}} \tag{6.36}$$

where OTT is the on-target time. The on-target time can be calculated from Eq. (6.31):

$$\begin{aligned} \text{OTT} &= \frac{1}{\dot{N}_p} \\ &= 4.88 \text{ E} - 6 \text{ s} \\ &= 5\ \mu\text{s} \end{aligned} \tag{6.37}$$

This will result in

$$\begin{aligned} B &= \frac{1}{5 \text{ E} - 6} \\ &= 2 \text{ E5} \\ &= 200 \text{ kHz} \end{aligned} \tag{6.38}$$

Note that the IFOV of the detector from Eq. (6.26) is about 60 μrad. This value is slightly larger than the diffraction limit of the 8-in lens:

$$\begin{aligned}\theta_d &= \frac{\lambda}{D}\\ &= \frac{10.6\,\mathrm{E}-6}{8 \times 2.54 \times \mathrm{E}-2}\\ &= 52\ \mu\mathrm{rad}\end{aligned} \tag{6.39}$$

where θ_d is the lens diffraction limit. This condition should be met in an acceptable design.

Using the given parameters in S/N ratio, Eq. (6.33), we can compute the S/N as a function of range R:

$$\begin{aligned}\mathrm{S/N} &= \frac{(20)(0.0324)(0.01)(0.40)}{\pi R^2 (6.6256\,\mathrm{E}-34)(2.83\,\mathrm{E}13)(1/0.50)(2\,\mathrm{E}5)}\\ &= \frac{1.10\,\mathrm{E}11}{R^2} \quad (R \text{ in m})\\ &= \frac{1.10\,\mathrm{E}5}{R^2} \quad (R \text{ in km})\end{aligned} \tag{6.40}$$

These values of S/N can be converted to dB units and plotted as a function of range, as shown in Figure 6.9. For example, at a range of 5 km:

$$\begin{aligned}S/N &= \frac{1.10\,\mathrm{E}5}{(5)^2}\\ &= 36.43\ \mathrm{dB}\end{aligned} \tag{6.41}$$

The straight line of Figure 6.9 shows the *free space* S/N versus range curve. These values can be modified for various weather condition losses occurring at transmitted frequencies. Approximate atmospheric losses at 10.6 μm wavelength for several nominal conditions are given in Table 6.1. Figure 6.9 also shows the S/N curves that include atmospheric losses computed with values from Table 6.1. Note that, at a given range, the two-way atmospheric loss is subtracted from the free space S/N ratio.

Since the laser beam covers the target, we may assume a glint target detection condition. From Figure 6.6(a), a 95 percent probability of detection, with false alarm probability of 10^{-7}, will require 14 dB of S/N

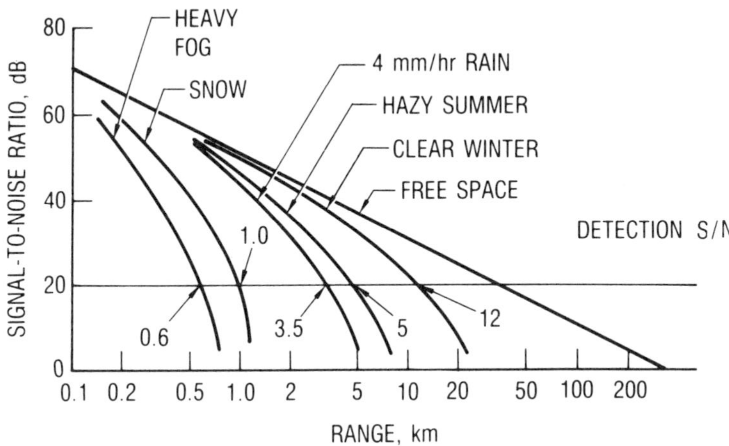

Figure 6.9 CO_2 Laser S/N *versus* range, including atmospheric losses for the imaging laser radar system.

Table 6.1 Atmospheric Attenuation Losses at 10.6 μm (from [3], page 340)

Condition	*Visibility* (km)	*Loss* (dB/km)
Clear winter day	12	0.4
Hazy summer day	5	1.7
Rainfall 4 mm/hr	3.5	3
Snow (1 mm/hr wet)	1	15
Heavy fog	0.6	30

ratio. Considering other unforeseen problems, let us require a 20 dB S/N for detection as given by the horizontal line of Figure 6.9. The intersection of this line with various curves of Figure 6.9 will give the detection range of the laser radar for the weather conditions under study. These values are summarized in Table 6.2.

The detection ranges of Table 6.2 happened to be identical to the visibility ranges of Table 6.1. This is not unusual for electro-optical systems such as thermal imagers and laser radars. The difference between TV-like optical systems and thermal imagers lies in the ability of thermal imagers, operating on temperature gradients, to work at night as well as during the day. The laser radar systems, having their own source of radiated power, can operate day or night without using temperature

Table 6.2 Summary of Detection Ranges of Imaging Laser Radar System

Condition	*Detection Range* (km)
Clear winter	12
Hazy summer	5
Rain (4 mm/hr)	3.5
Snow	1
Heavy fog	0.6

gradient or daylight. Laser radar systems, additionally, provide radar target range and range rate measurements.

From Figure 6.6(a), the false alarm probability of 10^{-7} will correspond to a threshold setting of 12 dB. This value is the 50 percent probability of detection S/N ratio.

Using Eq. (6.21), we find that the number of false alarms/s will be

$$\begin{aligned} N_{fa} &= \frac{P_n}{\text{OTT}} \\ &= 0.02 \end{aligned} \tag{6.42}$$

or 1 false alarm every 50 s per detector. If 15 receivers (detectors) are used, then the number of false alarms will be one every 3.3 s. As discussed earlier, this high false alarm rate can be reduced by considering consecutive detections in determining false alarms.

6.12 GROUND BACKSCATTERING AND INTERFERENCE SIGNALS

When transmitted laser pulses are intercepted by the ground, the S/N ratio can be computed by using the ground area as the radar target cross section σ. Referring to Figure 6.10, this area is calculated as follows:

$$\sigma = (R\theta_{\text{BW}})\left(\frac{c\tau}{2}\right) \tag{6.43}$$

where σ is the cross section of the laser radar target. The first term after the equal sign is the beam coverage on the ground. The second term is the equivalent range of a single laser pulse as shown in Figure 6.8. If this pulsewidth is greater than the laser beam coverage, then the term should

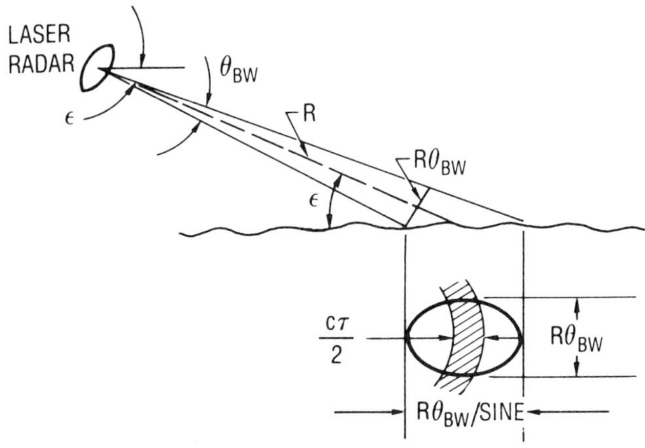

Figure 6.10 Laser radar beam interception with ground.

be replaced by the beam coverage on the ground. In this case, Eq. (6.43) becomes

$$\sigma = (R\theta_{BW})(R\theta_{BW}/\sin \epsilon) \tag{6.44}$$

where ϵ is the depression angle. Equation (6.44) approximates the elliptical area coverage of the laser beam on the ground. The laser radar cross section values of Eqs. (6.43) and (6.44), together with appropriate reflectivity characteristics, ρ, are used in the signal power equations (6.7) and (6.10). The value of the reflectance parameter, ρ, depends on the intercepted ground terrain (sand, cement, vegetation) as well as laser radar wavelength.

Interference signals, unlike the backscattered energy discussed earlier, are caused by natural radiation emittance of the sky, clouds, and ground. This radiation, in some cases, increases the noise level of the laser radar receiver. Natural radiation emittance usually is expressed in

$$\mathrm{W/(cm^2\ sr\ \mu m)} \tag{6.45}$$

Thus, the emitted radiation is per unit area of the radiating surface, with a distribution over a spherical angular coverage in sr and over a wavelength determined by the laser radar receiver.

For a given laser radar, the background noise power level can be expressed as

$$N_B = \left[\frac{\pi}{4}(R\theta_{\text{BW}})^2\right]\left[\frac{1}{R^2}\right]\left[\frac{\pi D^2}{4}\right] \cdot \int_{\lambda_1}^{\lambda_2} \text{W}(\lambda)d\lambda \tag{6.46}$$

where

N_B = background noise
D = lens diameter
θ_{BW} = lens beamwidth
R = laser background range
$\text{W}(\lambda)$ = spectral emittance, W/(cm² sr μm)

The first term of Eq. (6.46) represents the area of laser beam at range R. The product of the second and the third terms of Eq. (6.46) represents the equivalent lens solid angle in sr. The integral represents the background emission. Equation (6.46) can be simplified into the following form:

$$N_B = \frac{\pi^2}{16} D^2 \theta_{\text{BW}}^2 \int_{\lambda_1}^{\lambda_2} \text{W}(\lambda)\, d\lambda \tag{6.47}$$

The values of spectral emittance as a function of wavelength are measured and represented in Figures 6.11 and 6.12. Figure 6.11 gives the radiance of a cloudless sky as a function of wavelength. Figure 6.12 represents solar reflection from clouds and from earth, as well as self-emission from earth.

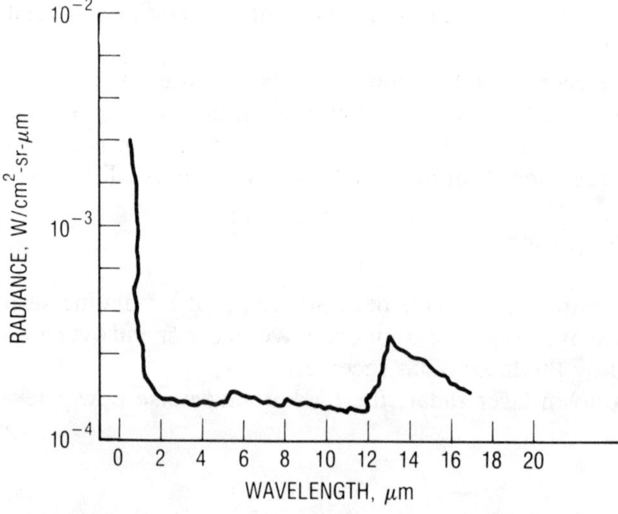

Figure 6.11 Sky background emission for a clear day.

Note from Figure 6.12 that solar reflection decreases exponentially with increasing wavelength. Self-emission from earth, however, increases with increasing wavelength. Since laser radar systems usually operate in a narrow frequency band, we can approximate the integral of Eq. (6.47) by assuming W(λ) to be constant over this frequency region:

$$N_B = \frac{\pi^2}{16} D^2 \theta_{BW}^2 W(\lambda) \Delta\lambda \tag{6.48}$$

where Δλ is the wavelength equivalent of the laser receiver bandwidth. This value can be obtained from the definition of wavelength:

$$\lambda = \frac{c}{f} \tag{6.49}$$

and

$$\begin{aligned} |\Delta\lambda| &= cB/f^2 \\ &= \lambda^2 B/c \end{aligned} \tag{6.50}$$

where B is the bandwidth of the laser radar receiver. Note that Eq. (6.50) gives the "equivalent wavelength" of receiver bandwidth B in a heterodyne receiver. In the case of direct detection Δλ is determined by the optical rejection filter.

As a numerical example of a background emittance calculation, consider the previously given laser radar with the following parameters for the system noise computation:

$$N = \frac{hf}{\eta} B \tag{6.51}$$

where

$h = 6.256\ \mathrm{E} - 34\ \mathrm{W} - \mathrm{s}^2$
$f = 2.83\ \mathrm{E}13\ \mathrm{Hz}$
$\eta = 0.50$
$B = 2\ \mathrm{E}5\ \mathrm{Hz}$

Using these values in Eq. (6.51), we get the noise level of the laser radar receiver

$N = 7.5\ \mathrm{E} - 15\ \mathrm{W}$

$$N = -141 \text{ dB} \tag{6.52}$$

Since this laser radar is designed for ground-level operation, let us compute the noise level caused by earth's self-emission. For this, we can compute the parameters of Eq. (6.48) as follows:

$$\begin{aligned} &D = 8 \text{ in}; \theta_{BW} = \lambda/D = 52\ \mu\text{rad}; \\ &\Delta\lambda = 7.5\text{ E} - 8\ \mu\text{m}; W(\lambda) = 1\text{ E} - 3\text{ W/(cm}^2\text{ sr }\mu\text{m)} \end{aligned} \tag{6.53}$$

where $\Delta\lambda$ is computed from Eq. (6.50) and $W(\lambda)$ is obtained from Figure 6.12. Using the parameters of Eq. (6.53) in Eq. (6.48), we get

$$\begin{aligned} N_B &= 5.16\text{ E} - 17\text{ W} \\ &= -162.8\text{ dB} \end{aligned} \tag{6.54}$$

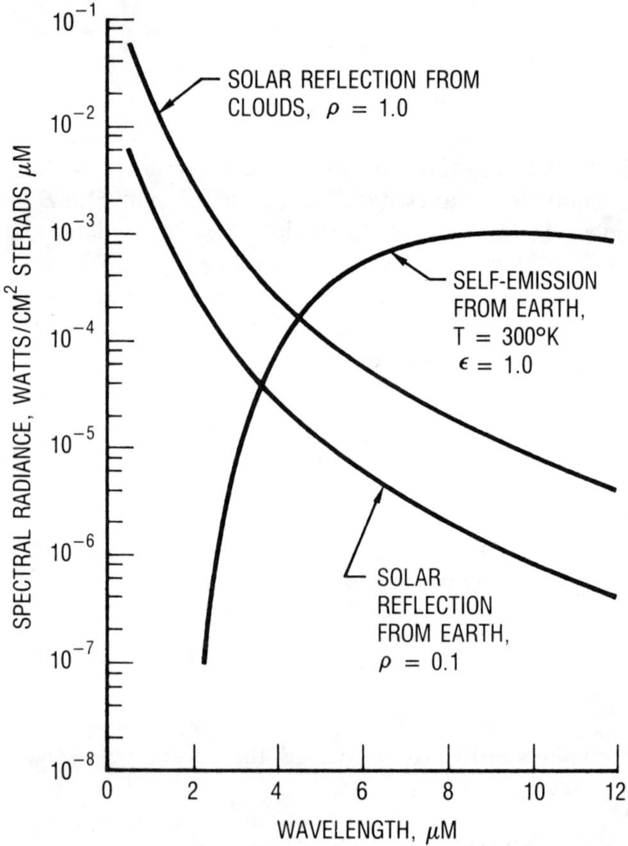

Figure 6.12 Solar reflections from clouds and earth and self-reflection from earth (from [1], page 86).

Note that this earth self-interference noise level is considerably lower than the system noise level of Eq. (6.52). Thus, it is not expected to increase the receive system noise level.

Note also that $\Delta\lambda$ of Eq. (6.50), and the subsequent computed value of Eq. (6.53), represent the bandpass equivalent of the laser receiver bandwidth. This is *not* the bandpass of the optical rejection filter, which usually has a bandpass of about 0.15 μm about the transmitted wavelength of 10.6 μm.

6.13 NUMERICAL EXAMPLE: SPACEBORNE LASER RADAR SYSTEM

We wish to design a spaceborne NdYAG laser radar system with a target-tracking range of 600 km. After several parameter iterations, we may arrive at the following parameter values:

$$\begin{aligned} \text{optics diameter} &= 20 \text{ cm} \\ \text{wavelength} &= 1.06\ \mu\text{m} \\ \text{peak pulse power} &= 2 \text{ MW} \\ \text{pulsewidth} &= 10 \text{ ns} \\ \text{target cross section} &= 2 \text{ m}^2 \\ \text{target reflectivity} &= 0.1 \\ \text{transmitted PRF} &= 100 \text{ Hz} \\ \text{tracking S/N} &= 3 \text{ dB} = 2 \\ \text{detectivity } D^* &= 10 \text{ E}10 \text{ cm Hz}^{1/2}/\text{W} \end{aligned}$$

Additional laser radar parameters can be computed or assumed using these data. The laser radar beamwidth can be computed

$$\begin{aligned} \theta_{BW} &= \frac{\lambda}{D} \\ &= 5.3\ \mu\text{rad} \end{aligned} \tag{6.55}$$

Using an $f/\#$ of 2, the focal length, f_1, becomes

$$\begin{aligned} f_1 &= (f/\#)D \\ &= 40 \text{ cm} \end{aligned} \tag{6.56}$$

When Eqs. (6.55) and (6.56) are used in Eq. (6.11), the detector area becomes

$$\begin{aligned} A_d &= f_1^2 (\theta_{BW})^2 \\ A_d &= 4.5 \text{ E} - 8 \text{ cm}^2 \end{aligned} \tag{6.57}$$

This is the area of a square detector 2 × 2 μm in size. But NdYAG detectors cannot be manufactured at these sizes. The manufactured detectors are orders of magnitude larger in size. Assuming a 250 × 250 μm detector, the detector area will be 6.25E − 4 cm².

The lens area can be written

$$A = \frac{\pi}{4} D^2$$

$$= 0.0314 \text{ m}^2 \tag{6.58}$$

The receiver bandwidth can be considered to be the inverse of pulse width as follows:

$$B = \frac{1}{\tau}$$

$$= 100 \text{ E6 Hz} \tag{6.59}$$

Using the parameters of the example problem in signal Eq. (6.7), we get

$$S = \frac{8(2\text{ E6})(0.0314)^2(2)(0.1)(0.5)}{\pi^3 R^4 (1.06\text{ E} - 6)^2}$$

$$= 4.5\text{ E13}/R^4 \quad (R \text{ in m}) \tag{6.60}$$

where an optical efficiency τ_o of 0.5 is assumed. The receiver noise can be computed from the detector noise level equation (Section 6.5) as follows:

$$N = [(6.25\text{ E} - 4)(100\text{ E6})]^{1/2}/10^{11}$$

$$= 2.5\text{ E} - 9\text{ W} \tag{6.61}$$

When Eqs. (6.60) and (6.61) are used, the S/N ratio becomes

$$\text{S/N} = \frac{1.8\text{ E22}}{R^4} \tag{6.62}$$

Note that the S/N ratio of Eq. (6.62) is computed on a per pulse basis. Integration of several pulses will increase this S/N proportional to the number of pulses integrated (neglecting integration loss). Assuming that the tracking loop is capable of integrating for 0.25 s and considering that 100 pulses are transmitted per second, the S/N improvement factor will be 25. Using this and a tracking S/N of 3 dB (factor of 2), we have

$$R^4 = \frac{(1.8\text{E} - 22)(25)}{2}\ \text{m}^4$$
$$R = 688\ \text{km} \tag{6.63}$$

This tracking range exceeds the original requirement of 600 km.

From the given values, a number of other important laser radar parameters can be computed. The PRF of 100 Hz will correspond to an interpulse period T and an unambiguous R_a of

$$T = \frac{1}{\text{PRF}}$$
$$= \frac{2R_a}{c} \tag{6.64}$$

The unambiguous range is defined as the range where target pulse return comes within the first interpulse period. From Eq. (6.64), we get

$$R_a = 1500\ \text{km} \tag{6.65}$$

This value exceeds the design range of 600 km as it should. The beamwidth of the laser radar, Eq. (6.55), at the detection range of 600 km will result in a linear dimension of

$$l = R\theta_{\text{BW}}$$
$$= 3.18\ \text{m} \tag{6.66}$$

or an approximate cross-sectional area of 10 m^2. This value exceeds the target cross section of 2 m^2, justifying the point target assumption. The pulse width τ of 10 ns is equivalent to a resolution range of

$$\Delta R = \frac{c\tau}{2}$$
$$= 1.5\ \text{m} \tag{6.67}$$

These values are within the limits of a target cross section of 2 m^2, which was given in the design of the laser radar.

Background interference can be computed from Eq. (6.48) and the data of Figure 6.12. This computation will show that the interference is below the system noise level as given by Eq. (6.61).

REFERENCES

1. C. G. Bachman, *Laser Radar Systems and Techniques*, Artech House, Dedham, MA, 1979.
2. C. T. Due and L. M. Peterson, *Optical-Mechanical Active/Passive Imaging Systems*, Volume 1, Environmental Research Institute of Michigan, May 1982.
3. A. V. Jelalian, "Laser Radar Theory and Technology," in E. Brookner (ed.), *Radar Technology*, Artech House, 1977, Dedham, MA, p. 344.
4. J. H. Shapiro, R. W. Reinhold, and D. Park, "Performance Analysis of Peak Detecting Laser Radars," *Proc. SPIE*, Vol. 663, 1986, pp. 38–56.
5. S. A. Hovanessian, *Radar System Design and Analysis*, Artech House, Dedham, MA, 1984.
6. D. M. Papurt, J. H. Shapiro, and S. T. Lau, "Measured Turbulence and Speckle Effects in Laser Radar Target Return," *Proc. SPIE*, Vol. 415, 1983, pp. 166–178.
7. W. L. Wolfe and G. J. Zissis, *The Infrared Handbook*, Environmental Research Institute of Michigan, Ann Arbor, 1978, pp. 22–87.
8. K. F. Hulme et al., "A CO_2 Laser Rangefinder Using Heterodyne Detection and Chirp Pulse Compression," *Optical and Quantum Electronics*, Vol. 13, 1981, pp. 35–45.

Chapter 7
Multiple Sensor Tracking and Data Fusion

by Samuel S. Blackman

The primary purpose of a sensor fusion consideration is to complement the data of one sensor with that of another sensor in order to obtain better target position information, to eliminate false alarms, and to reduce susceptibility to countermeasures. For example, consider the case of a microwave radar used for large area search with a laser radar for improved angular position accuracy. The microwave radar, because of its large beamwidth, can search a large volume of space in a short time and obtain angular position and range. Subsequently, the laser radar can be pointed in the direction of the target given by the microwave radar. After a small area search the laser radar can lock onto the target and track it. This will improve target angular position accuracy as the laser beam is considerably smaller. Microwave radar can continue its search designating targets for laser tracking.

The laser radar can also confirm the existence of the target and assure that the microwave radar target was not a false alarm. Additionally, since microwave and laser radars operate in different frequency bands, electronic countermeasures used against one will be ineffective against the other. This, of course, will reduce susceptibility to countermeasures.

Additional sensor fusion candidates might be passive IRST–laser radar, thermal imager–laser radar, millimeter wave–thermal imager, or microwave radar–thermal imager. In each of these cases, the sensors are complementary. The passive IRST system can be used for large area search. After locating the target the laser radar (laser range finder) can be used to find target range.

In the second case, the thermal imager is used for search and target recognition and the laser radar is used to obtain target range. In the third example of a tactical short-range application, the millimeter wave radar is used for target search and the thermal imager is used for target recognition. The fourth example is similar to the third, with the microwave radar replacing the MMW radar, for a larger area search.

A recent example of a multiple sensor tactical system is the low altitude navigation and targeting for night (Lantirn) manufactured by Martin Marietta Corporation and discussed in detail in *Aviation Week and Space Technology* (April 25, 1988). This system, which consists of a navigation pod and a targeting pod, can be mounted on F-16 fighters. The sensor systems of the navigation pod consist of a *terrain following radar* (TFR) and a *forward looking infrared radar* (FLIR) system. The targeting pod includes a FLIR and a laser radar system. To make the system operational, hundreds of flight test hours were spent for harmonizing the operation of the FLIR, TFR, navigation system, and weapon delivery system.

This chapter gives an overview of the issues and methods involved in the use of multiple sensors for target surveillance and tracking. First, we discuss the problem of data association where sensor observations are distributed among target tracks. We then discuss techniques for estimation of target kinematics, such as position and velocity, and target attributes, such as type and shape. We conclude with a discussion of proposed sensor allocation or sensor management for a given target surveillance situation.

7.1 INTRODUCTION

Many benefits may be derived from the use of multiple sensors in a target surveillance system, such as the potential for broader coverage and the use of complementary information, such as accurate angular position of electro-optical systems and range and range rate information of radar systems, to provide improved tracking. Resistance to countermeasures can also be provided by multiple sensors through their operating modes, such as active for radars and passive for electro-optical systems and a wide range of sensor operating wavelengths (frequencies). Since most electronic countermeasure devices, such as jammers, operate at specific frequencies, all of the sensors of a multiple sensor system will not be degraded in the presence of these jammers.

In this chapter we present a survey of the issues and methods for associating and fusing data from multiple sensor systems into target tracks. As exemplified by the discussion of previous chapters, there is a rather well-developed theory for design and analysis of individual sensors.

On the other hand, the methods required to efficiently utilize the data from multiple sensors for such tasks as tracking and identifying targets are not nearly as well developed. Thus, the goal of this chapter will be to present a broad coverage of the many methods available. The mathematical details will not be presented but an extensive list of references will be provided for the reader who wants more detailed information.

The chapter will cover three broad areas. The first, data association, refers to partitioning the sensor data according to source. During this process, sensor observations determined to originate from the same target source are used to form target tracks. Once tracks are formed, we can estimate the number of targets present and such characteristics as position, velocity, and classification. These allow higher order inferences, such as target intent.

The chapter begins with an overview of methods in common use for data association and then it addresses the important issue of the level of data association. The basic alternatives here are whether to partition each sensor's data individually into sensor-level tracks and then combine tracks or to perform track formation directly from the measurements of all the sensors.

The second broad category to be discussed in this chapter is the estimation of target characteristics once data association has been performed. This estimation process involves target kinematics, such as position and velocity, as well as target attributes, such as target types and cross section. An important goal of attribute estimation is target identification. Combined with target kinematics, this can be used to anticipate future target behavior. Finally, we discuss the unique problems and opportunities that arise when sensor systems are physically separated (or distributed).

The third general topic presented is that of sensor allocation. This important topic has only recently begun to be addressed theoretically, rather than in an ad-hoc manner. The chapter closes with an assessment of present technology and a prediction of future trends.

7.2 DATA ASSOCIATION

Data association, with the goal of partitioning observations into tracks, is the key function of any target surveillance system (References [1] and [2]). We begin this section with an overview of the elements of data association as exemplified by the recursive *multiple-target tracking* (MTT) system shown in Figure 7.1.

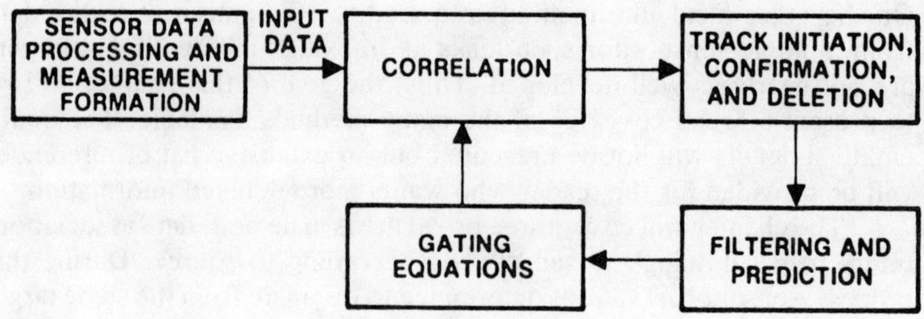

Figure 7.1 Elements of multiple target tracking system.

The sensor measurement and data formation function has been the subject of previous chapters of this book and is generally not considered part of data association. However, as discussed in [2], there are many opportunities for feedback between tracking and measurement functions. An example is the use of adaptive thresholding within the region of target return expected from the target tracking system. Another example is the use of tracking information to help resolve targets that are unresolved because of close spacing in the sensor measurement.

Referring to Figure 7.1, assume the existence of previously formed tracks. Incoming observations are considered for the update of the existing tracks in the correlation block. Gating tests determine which observation-to-track pairings are "reasonable," and a more refined correlation algorithm is used to determine final pairings. Observations not assigned to existing tracks can initiate new, tentative tracks. A tentative track becomes confirmed when the number and quality of the observations included in the track satisfy confirmation criteria. Similarly, low quality tracks, usually determined by the update history, are deleted. Finally, after including the new observations, tracks are predicted ahead to the arrival time for the next set of observations. Gates are placed around these predicted positions and the processing cycle repeats. Next we will discuss in more detail several philosophies commonly employed for data association. See also the track-while-scan discussion of Chapter 2.

7.2.1 Data Association Methodology

The nearest-neighbor approach to data association determines a unique pairing so that at most one observation can be paired with a given track. The method is based upon likelihood theory with the goal of minimizing the distance from the observation to the predicted position of the target.

The alternative is the all-neighbor approach, described in Reference [3], which incorporates all observations within a gated region about the predicted track position into the update of the track. In this case, a given observation can be used to update more than one target track. Track update is then based on a probabilistically determined weighted sum of all observations within the track gate.

Another important consideration is the timing of data processing. An ideal situation will be to process all observations from all time with a batch processor. Since this is not computationally feasible, the most common approach has been to perform processing in a recursive (or sequential) manner as data are received. For example, a set of observations from a time frame (or scan) of a scanning radar may be collected and data association performed on these observations during the time period when the sensor is collecting the data from the next scan or radar frame. Using this approach, once data association decisions are made, they are irrevocable.

A deferred decision approach to data association, as exemplified by the branching algorithm of Reference [4] and the more structured multiple hypotheses tracking method of Reference [5], allows the final decision on difficult data association situations to be postponed until receipt of more information, such as the next frame of data. In effect, this approach allows a modified version of batch processing. Alternative hypotheses are formed and evaluated when later data are received.

7.2.2 Data Association for Multiple Sensor Systems

The use of multiple sensors has the potential, through the inclusion of more varied data, to greatly improve data association. For example, the combined use of accurate range data from a microwave radar sensor and accurate angular information from an *infrared* (IR) sensor can lead to accurate target track position estimates, reducing data association uncertainty. However, the problem of initially combining radar and IR data into a target track when the IR data have no range information is difficult. This difficulty may be further compounded by the resolution differences of IR and radar sensors. Infrared sensors can resolve closely spaced multiple targets that appear as a single target to a radar sensor. Reference [6] demonstrates an approach to data association for a representative multiple sensor system.

Input data from a variety of sensors allow introducing target attributes (target type, size) into the association logic. This will require a more general association logic in which both the physical position parameters and target attributes are used simultaneously. Reference [2] discusses some aspects of these issues that currently are being studied.

7.3 LEVEL OF DATA ASSOCIATION: CENTRAL LEVEL *VERSUS* SENSOR LEVEL TRACKING

The first issue in developing a multiple sensor surveillance system is at the level of data association. Data can be associated in each sensor or can be combined from all sensors.

Figure 7.2 illustrates sensor-level tracking. Target tracks in this case are established primarily on measurements received from the individual sensors, but some communication among the sensors and between the sensors and the central track file may be used to update sensor-level track files. However, the sensor-level tracks must eventually be combined into a central track file. Thus, under this approach each sensor would have separate track files, and a central track file would be formed as a composite.

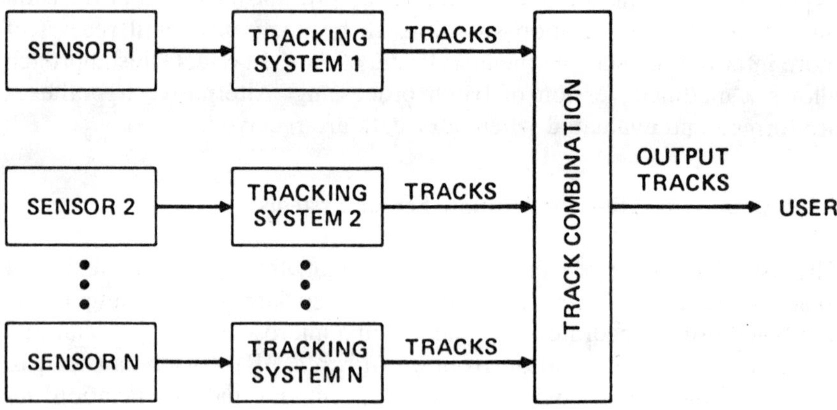

Figure 7.2 Sensor level tracking.

Sensor-level tracking is used to reduce data-bus loading and computational loading (in any single processor), and its distributed tracking capabilities enhance survivability. Certain computational advantages may result from the parallel processing possible in this approach. Also, if one sensor becomes degraded, its observations will not affect the sensor-level tracks of others. Finally, sensor-level tracking allows for a filter design that is tailored specifically to the individual sensors.

A major difficulty associated with the use of sensor-level tracking arises when the sensor tracks ultimately are combined. First, we must define an extensive track-to-track association logic. This may be difficult if we are using deferred decision data association logic, because multiple hypotheses logic must be manipulated at both sensor and combined tracking levels. Finally, even if we perform correct track association, the combining logic must account for common error sources due to potential target maneuvers. Reference [7] suggests a practical solution for this problem through adjusting the Kalman filter covariance matrix after track combining.

The alternative to maintaining sensor-level tracks is for all sensor data to be sent directly to a central processor, where a master track file is maintained. This approach, illustrated in Figure 7.3, has a number of advantages. First, more accurate tracking should be possible if all data are processed at the same place. A target track that consists of observations from more than one sensor should be more accurate than the tracks established on the partial data received by the individual sensors. Thus, the central processing track approach should lead to fewer misassociations. Second, processing sensor reports directly avoids the difficulties associated with combining sensor-level tracks.

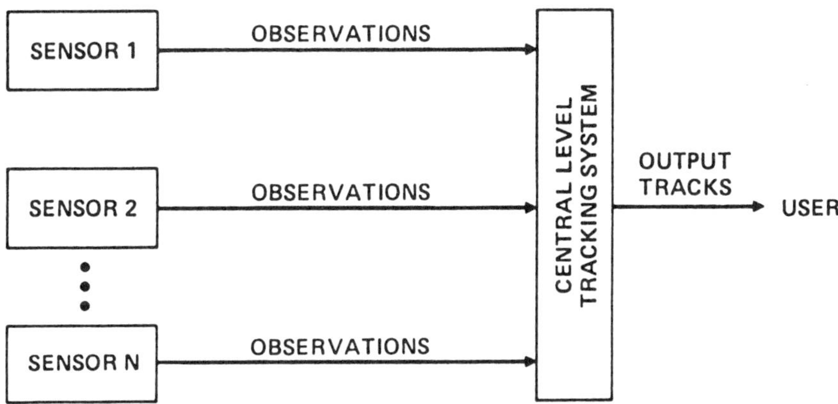

Figure 7.3 Central level tracking.

The central-level tracking approach suffers primarily because of the amount of observation data that must be transferred. Note that the periodic transmission of tracks, in the sensor level case, required much less data transfer than if all of the observations forming the tracks were transferred. Other potential problems with central tracking relate to its military uses, where vulnerability to central level failure (or destruction) are important.

Since both the sensor and central level approaches suffer from potential problems, a combined approach [8] may be proposed, as shown in Figure 7.4, where both central and sensor level tracks are maintained. One technique is to form central level tracks from sensor level tracks initially and then to send selected observations to the central level tracks for track update. Another potential saving precombines observations from different sensors, but the same target, before inclusion in the processing.

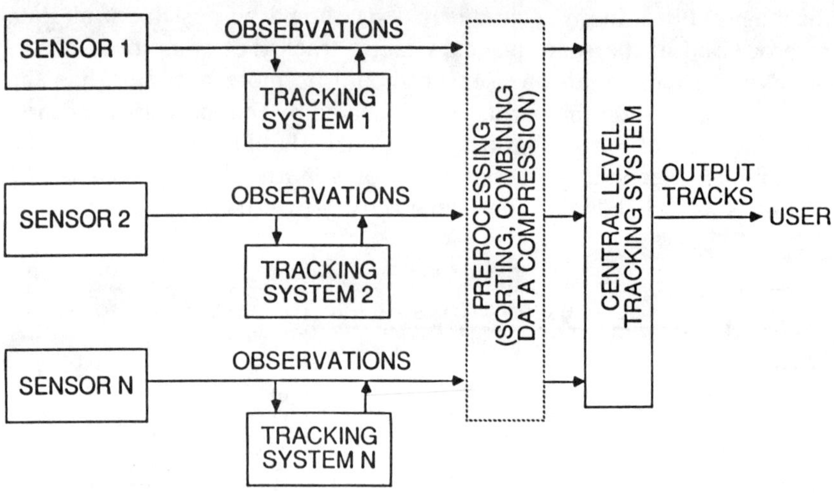

Figure 7.4 Combined sensor and central level tracking approach.

7.4 POSITION FILTERING AND PREDICTION

The purpose of the kinematic or position tracking filter is to update the track state estimate each time new observation data are received and to form future state estimates. The track state usually consists of the target position, such as range and angular values of azimuth and elevation, and velocity information. Kinematic filters, in effect, provide a weighting for the present estimate of target position and the received observation. This weighting is determined by considering the present position accuracy and the estimated error of the obtained observation. Future target positions are estimated through the use of these "weights."

7.4.1 Kalman *versus* Suboptimal Filtering

Kalman filtering [9] provides a convenient means for determining the weightings (denoted as gains) to be given to input measurement data. It also provides an estimate of the target tracking error statistics, through the covariance matrix. The Kalman filter covariance matrix can then be used for gating, along with measurement error statistics. These gates determine the future feasibility of observation-to-track pairing and also the final observation-to-track assignment.

For some applications it may be desirable to use approximations to the Kalman filter, in which the gains are predetermined and *a priori* estimates of track statistics are used for gating and data association (suboptimal filtering). However, when considering multiple sensor applications, and in particular when a variety of dynamic targets are to be tracked, the flexibility and the adaptability of the Kalman filter are essential. The Kalman filter chooses the gain sequence and estimates the track accuracy in accordance with the variations (in terms of accuracy and update rate) of input data and the actual target geometry. *A priori* estimates required for fixed gain filtering are hard to determine, especially for maneuvering targets. Experience shows that inaccurate estimates lead to significant performance degradation for fixed gain filters.

7.4.2 Passive (Angle Only) Sensor Tracking

The design of tracking filters for radar data is relatively straightforward, because the radar, through range and angular measurements, provides a three-dimensional target position. However, passive sensors such as *infrared search and track* sets (IRST) and *radar warning receivers* (RWR) only provide angular measurements. Thus, the full three-dimensional target position information will not be available with a single passive sensor.

Passive ranging techniques are available for target range calculations using multiple angular observations and triangulation methods. These methods use simple geometric relations to calculate range, from angular target position, information obtained from two different locations. The distance between the observation points should be known. Target position accuracy thus obtained will depend primarily on the accuracy of the filtered angular data. However, as we will discuss later and in Reference [10], a complex logic may be required in order to eliminate false targets (ghosts) formed by triangulation in the presence of multiple targets.

One approach to tracking with a single passive sensor is to be content with merely angle information and, for example, just track angle and angle rate. Alternatively, it may be possible to estimate a full three-dimensional target state vector if there is an own ship (sensor bearing platform) maneuver and if the target does not maneuver [11].

7.4.3 Combining Active and Passive Sensor Data

In a number of applications, it is often necessary to restrict active transmission in order to avoid alerting the enemy to one's presence and position. One approach to this problem is to limit active sensor transmission to providing only sporadic active sensor data to augment the passive data. A typical application of this approach would use a microwave radar to aid in a system where an IRST system is the primary sensor for track initiation and maintenance.

The manner in which sporadic radar data can be used in conjunction with an IRST system is illustrated in the simulation results of Figure 7.5. This figure shows the mean angular tracking error in response to a target that performs a 3 g maneuver for 12 s. As described in [2], Chapter 3, the angular tracking filter chosen for this application required the use of range and range rate in the Kalman filter transition matrix. Thus, there is a coupling from the range filter into the angle filter. Then, as range tracking performance degrades with the decreasing rate of active radar data, angle tracking will also suffer.

The IRST angular measurement data were received at two second sampling intervals while the update rate of radar range and range rate data were varied from 2 to 10 to 20 s. Typical IRST and radar measurement parameters were assumed for this simulation.

Note that, in Figure 7.5, there is a considerable degradation when radar data are restricted to 20 s update intervals, but going from 2 to 20 s intervals has much less effect. Also, it should be noted that these results are for short ranges of less than 10 nmi. For longer ranges, the decreased rate of radar data does not degrade angular target tracking to a great extent. Thus, the general approach of using a limited amount of active sensor transmission to aid a primarily passive sensor tracking system seems promising for future applications.

7.5 FUSION OF MULTIPLE SENSOR ATTRIBUTE AND TARGET TYPE DATA

Future multisensor systems will provide a variety of target attribute data, and a major current research area is the development of techniques to combine (or fuse) these data. Typically, the ultimate goal of this fusion process is to determine target type and identity.

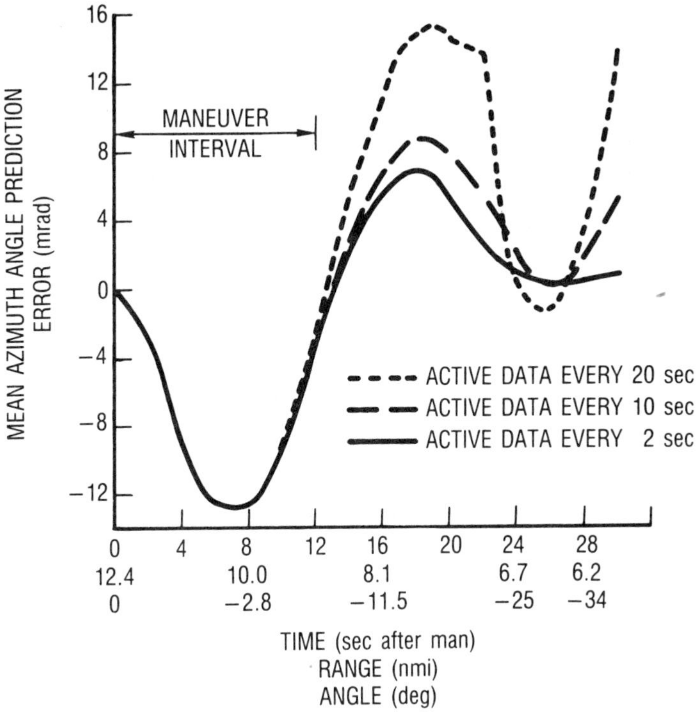

Figure 7.5 Tracking with combined passive and active sensor data.

Pattern recognition is one method of using multiple sensor data to identify targets. Using this approach, we would determine the appropriate set of features to be formed from multiple sensor observation data and the best weighting (confidence level) for these data. This, however, is a complex process and probably is only feasible for a limited number of sensors. Also, we would have to redefine the features and weightings when another sensor is added to the system.

The approach we will discuss in this chapter assumes that each sensor first processes its own data, then produces its current best estimate of target attributes. We also assume that the confidence associated with each output to be transmitted or to be known from previous experience. Thus, the problem becomes one of combining sensor attribute data to specify target types and the associated confidence level.

As an example of multiple sensor attribute data, Figure 7.6 shows the information that may be available to an airborne interceptor system using multiple sensors and advanced processing methods. This target attribute information may include the type of engine, type of radar, target

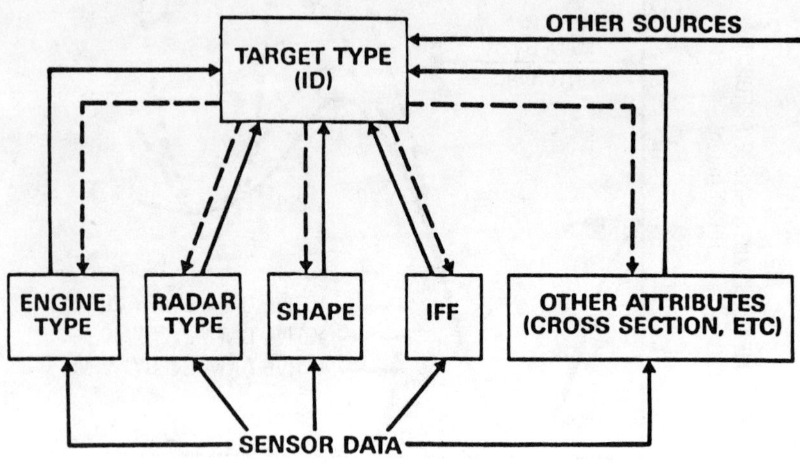

Figure 7.6 Target identification information flow diagram.

shape, response (or lack of) to *interrogation friend or foe* (IFF), and target RCS. Finally, other information, such as flight path characteristics and possible intended target destination, may also be added from other sources.

Table 7.1 shows potential attributes of typical aircraft type targets, assuming that four types of engines, three types of radars, and three target shapes can be categorized. The table gives the estimated attributes of four targets obtained from multisensor data together with actual target characteristics. Note that in a complete mechanization of this procedure a confidence level will be associated with each estimated attribute. This confidence level can then be used in calculating the overall confidence level of target identification. Also, more than one attribute type can be associated with a given target type. For example, a given aircraft may carry several types of radar.

From Table 7.1 we note that all of the attributes of targets A and D match with their actual characteristics. In the other targets, only two out of three attributes are matched. From this table we can infer that there may be considerable ambiguity with only a single measured attribute, and two or more attribute measurements will substantially reduce the level of ambiguity.

Table 7.1 Representative Aircraft Type Targets Together with Estimated Attributes

Target Type	*Engine Type*	*Radar Type*	*Shape Category*	*Actual Target*
A	E4	R2	S3	E4/R2/S3
B	E1	R2	S1	E3/R2/S1
C	E4	R1	S3	E1/R1/S3
D	E4	R3	S2	E4/R3/S2

The fusion of multiple sensor data is complicated by the uncertainties that exist in the measurements and the attribute assignment process. Furthermore, uncertainties may also exist in the initial definitions of target characteristics. In the next section, we will discuss three approaches to solving this problem.

7.5.1 Possibility Theory

The application of possibility theory to the sensor fusion problem starts by assigning a unit possibility to all possible hypotheses or, in our case, all possible target characteristic types. The attribute estimates are related to target types through confidence branches. For example, if a given estimated attribute matches the characteristic of the target type, the possibility of that target type is unaltered. If an estimated attribute is inconsistent with a given target characteristic type, the possibility of that target type is multiplied by $(1 - C)$, where C is the confidence level of attribute estimate.

As an example of possibility theory application consider Target A of Table 7.1 with the following attribute confidence levels: E4 (80 percent); R2 (90 percent); S2 (30 percent). (Note that shape category S2 is different from the one given in the table.) Using these values together with the actual target characteristics of E4/R2/S3, the resulting possibility measure is

$$P = (1)(1)(1 - 0.30) = 70 \text{ percent}$$

7.5.2 Bayesian Approach

The application of a Bayesian approach to the attribute and target identification problem requires *a priori* information and conditional probabilities. First the measurement process is defined by the relationship

$$P(x_m/x) = \text{probability of receiving measurement } x_m \text{ given that the true quantity is } x$$

which is assumed to be known. Then whenever measurement data are received the updated probabilities can be computed using Bayes' rule:

$$P(x/x_m) = \frac{P(x_m/x)P(x)}{P(x_m)} \tag{7.1}$$

where

$$P(x) = \text{prior probability associated with } x$$
$$P(x_m) = \sum_x P(x_m/x)P(x)$$

The process continues as $P(x/x_m)$ becomes the new prior probability for use when further data are received. To initiate the process, before any measurements are received, requires an initial estimate of the probability of x. For a numerical example of the Bayesian approach the reader is referred to Reference [12].

To summarize, Bayes' rule (7.1) is solved recursively as new data are received. This relatively simple relationship provides the estimated probabilities of target attribute based directly on the measurements x_m of involved quantities. However, as we will outline later and discussed in detail in [2], the estimated probabilities can be improved using known interrelationships, as expressed by conditional probabilities, between attributes and target types.

Target type estimates can be used to update target attribute estimates and *vice versa*. This can be accomplished because certain types of targets use specific types of equipment. For example, the radars used in passenger aircraft are different than the radars used in military fighters. In order to relate attribute and target type estimation, it is first necessary to specify conditional probability relationships between expected attributes and target types. Then the problem of jointly estimating target type and attribute values can be formulated as a special case of a general inference net, as discussed in detail in [2].

7.5.3 Dempster-Shafer Evidential Reasoning

Dempster and Shafer (References [12] and [13]) have developed a method that generalizes Bayesian inference, and much recent interest has arisen in application of this approach. The Dempster-Shafer approach has a number of unique features and advantages that are probably best illustrated by a numerical example using the information from Table 7.1.

Assume that an imaging *electro-optical* (EO) sensor declares a given target to be of shape category S3 and, furthermore, assume a confidence of 0.90 on the declaration. Referring to Table 7.1, note that both targets A and C are in shape category S3. Thus, corresponding to the confidence of 0.90 we assign probability mass value of 0.90 to the disjunction of A and C. Note that this approach differs from the Bayesian, in which probability must be assigned to individual propositions (0.45 to A and 0.45 to C). Finally, a mass value of 0.10, corresponding to 1 minus the confidence of 0.90, is assigned to θ, which represents the disjunction of all propositions (in this case all four target types). Thus, the result of the first sensor measurement is a mass assignment vector as defined

$$m_1 = \begin{bmatrix} m_1\,(A v C) & = 0.90 \\ m_1\,(\theta) & = 0.10 \end{bmatrix} \tag{7.2}$$

where A and C are target types. Next, assume that an electromagnetic-type sensor system determines the radar type to be R2, with a confidence level of 0.80. Since this radar type is associated with targets A and B, the mass assignment vector of this sensor will be

$$m_2 = \begin{bmatrix} m_2\,(A v B) & = 0.80 \\ m_2(\theta) & = 0.20 \end{bmatrix} \tag{7.3}$$

In order to combine the information given by Eqs. (7.2) and (7.3), Dempster's rule, shown in Figure 7.7, is used. The resulting mass vector is

$$m = \begin{bmatrix} m(A) & = 0.72 \\ m(A v B) & = 0.08 \\ m(A v C) & = 0.18 \\ m(\theta) & = 0.02 \end{bmatrix} \tag{7.4}$$

$m(AvC) = 0.90$ $m(\theta) = 0.10$	$m(A) = 0.72$ $m(AvB) = 0.08$	$m(AvC) = 0.18$ $m(\theta) = 0.02$
	$m(AvB) = 0.08$	$m(\theta) = 0.20$

Figure 7.7 Application of Dempster's rule to the numerical example.

In Figure 7.7, Dempster's rule is implemented by forming a matrix with the probability mass assignments to be combined given along the first column and the last row. Then, the computed elements (for a given row and column) of the matrix are the products of the probability mass values in the same row of the first column and the same column of the last row. For example, for the (1,1) element of the matrix shown in Figure 7.7 we have

$$m(A) = m(AvC)m(AvB) = (0.90)(0.8) = 0.72$$

Given propositions a_i, the assignment of elements to the mass vector resulting from Dempster's rule is according to the following principles:

1. The product of mass assignments to two propositions that are consistent leads to an assignment to another proposition contained within the two original propositions. For example:

$$m_1(a_i)m_2(a_i) = m(a_i)$$
$$m_1(a_1va_3)m_2(a_3va_4) = m(a_3)$$

2. Multiplying the mass assignment to uncertainty by the mass assignment to any other proposition leads to a contribution to that proposition:

$$m_1(\theta)m_2(a_3va_4) = m(a_3va_4)$$

3. Multiplying uncertainty by uncertainty leads to a new assignment to uncertainty:

$$m_1(\theta)m_2(\theta) = m(\theta)$$

This example illustrates Dempster's rule for the condition where there is no inconsistency (or assignment to a null hypothesis) between the knowledge sources. Inconsistency occurs, for example, when one knowledge source assigns mass to $a_2[m_1(a_2)]$ while a second assigns mass to $a_1[m_2(a_1)]$. The product of these mass values is assigned to a measure of inconsistency, k, of the form

$$m_1(a_2)m_2(a_1) = k$$

Relating to the present example, this inconsistency will occur if an IFF sensor produced a "friendly" response from a target for which there is probability mass associated with the enemy. Given that inconsistency between two knowledge sources occurs, it is necessary to perform a normalization such as to include this effect in the mass matrix, as discussed in [2].

A final interesting point associated with the Dempster-Shafer method is the concept of a probability interval defined by the support and plausibility for a given proposition. The support for a proposition is the sum of all the masses assigned directly to that proposition. For example, from Eq. (7.4), the support for A is 0.72. The plausibility of a given proposition is the sum of all masses not assigned to its negation. For example, the plausibility of A is 1.0 because all propositions represented in Eq. (7.4) contain A. On the other hand, the plausibility associated with C is

$$\text{Pls}(C) = 1 - m(A) - m(A \vee B) = 0.20$$

Since there is no mass assigned directly to C, the support of C is zero.

7.6 DISTRIBUTED SENSOR SYSTEMS

The discussion so far has been basically for local sensor fusion, which can be defined as the integration of data from sensors onboard a single platform. Next, we briefly consider the issues involved in the fusion of data from sensors that are spatially distributed so that they see the targets from different locations and viewing angles.

An important consideration for a distributed sensor system is the communication involved in transfer of information between the sensors. This can lead to a trade-off between allocating system resources for computation *versus* communication. Also, the potential cost of transferring all observations and the desirability of performing local data processing tends to favor the sensor level tracking approach. However, as we will discuss, techniques have been proposed in which selected observations are passed between sensor sites, so that central level tracking accuracy can be achieved by distributed data processing.

Another important issue is that of sensor control. Here, based upon survivability considerations, it is desirable that each individual sensor system have the capability to operate autonomously. This implies that, based upon locally available information, each site will determine where its sensors should point and what targets should be tracked.

Another important question involves track maintenance. Should all tracks be maintained at all sites or should each site maintain only a selected subset of target tracks? One common approach to this problem has each site maintaining two sets of track files. The first (global) track file contains tracks on all targets, with track information being communicated from other sites. The second (local) track file contains tracks on those targets that are being actively tracked by that site. The maintenance of a global track file at each site may be appropriate for internetted interceptor and ground defense radar systems but may not be feasible, due to the potentially enormous number of targets.

7.6.1 Example Approaches

We next outline three approaches to distributed sensor system design. All approaches maintain two sets of track files. The first approach [14] maintains local and global track files as described earlier, and each target is tracked cooperatively by a few selected radar sites. A complex logic is required to determine which sites maintain the responsibility for which targets and to ensure that the appropriate target observations are sent to the sites tracking the target.

For the first approach, in which selected observations are communicated, the observations are properly routed by comparing them with tracks in both the local and global track files. Given that an observation is a potential for track update, it may be used locally or sent to other sites actively maintaining that track. Thus, each site must know which other sites are tracking which targets. Observations not associated with local tracks or communicated to other sites become potential new tracks. Finally, based upon locally available information, each site must determine which tracks it should actively maintain and when it should assume or relinquish tracking responsibility.

A second approach [15] utilizes extensive track fusion. In order to perform this fusion a typical system will require the maintenance of a local and a common track file. The common track file contains information believed known to all sites. The local track file includes the common track file information for those targets tracked locally and the information derived from the local site sensors but not as yet transmitted to the other sites.

Each site will periodically transmit its local track file information so that other sites can update their track files (remember that in this case a local track file contains all local track information plus all information received from other sites). However, in order that the other sites can properly use this newly broadcast information, it is necessary to extract that portion previously distributed. Thus, the common track file provides a reference whereby a site can extract the new information and, in effect, disregard information that has previously been included.

A third and a simpler approach has each site maintain a local track file based only upon its own sensor observations. A global track file is also maintained as a composite of the best local tracks from all of the sites included in the system. Thus, whenever a given site determines that one of its local tracks is of higher quality than the corresponding global track, it replaces the global track with the local track and transmits the local track to other sites for use in their global track files.

To summarize, the first two approaches will be more accurate because, in effect, a track may use observations from more than one site. In particular, the first approach should be the most accurate because observation data are used directly, while the second approach relies upon track fusion. The third approach is the simplest but will be the least accurate, because any given track will be formed using only the observations from a single sensor site. However, since the best track for any given target is chosen for the global track file, the simplicity of this approach may make it preferable for systems in which the overlapping coverage is small or where the tracking requirements are not stringent.

7.6.2 Combining Passive (Angle Only) Tracks

An important advantage of distributed passive sensor systems is the ability to combine two angle-only tracks to form a full three-dimensional (range, azimuth, and elevation angles) track. This process, commonly referred to as *triangulation*, is illustrated in Figure 7.8. Here, the heading angles directed along the vectors $\overline{V}_1$ and $\overline{V}_2$ represent angle tracks from sensor sites S_1 and S_2. The two sites are separated by vector $\overline{M}$. The appropriate equations for the calculations are

$$\begin{aligned} \overline{V}_1 &= (\Lambda_{N1}\, 1_N + \Lambda_{E1}\, 1_E + \Lambda_{D1}\, 1_D)\, R_1 \\ \overline{V}_2 &= (\Lambda_{N_2}\, 1_N + \Lambda_{E2}\, 1_E + \Lambda_{D2}\, 1_D)\, R_2 + \overline{M} \end{aligned} \quad (7.5)$$

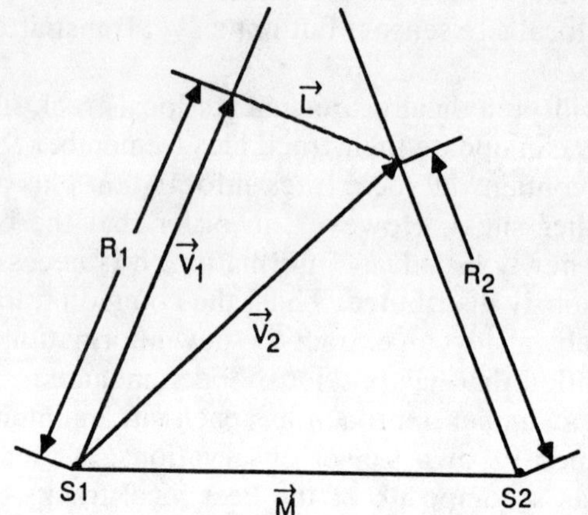

Figure 7.8 Triangulation method for passive sensor system.

where Λ_N, Λ_E and Λ_D represent the north, east, and down direction cosines and 1_N, 1_E, and 1_D are unit vectors in these directions. The goal is to solve for R_1 and R_2 that minimize the distance

$$D^2 = |\overline{V}_1 - \overline{V}_2|^2 \tag{7.6}$$

These equations can be solved for R_1 and R_2 satisfying the equation

$$\frac{\partial D^2}{\partial R_1} = \frac{\partial D^2}{\partial R_2} = 0 \tag{7.7}$$

Figure 7.9 gives range estimation standard deviation error as a function of the azimuth and elevation angular errors (assumed equal). The geometry of the encounter is also shown in the figure. Results are given for two separation distances ($M = 0.5$ and 1.0 nmi) and for two target ranges ($R = 25$ and 40 nmi). As expected, the range estimation error decreases with decreasing range and increasing sensor separation.

As illustrated in Figure 7.10, a major problem that can result from triangulation is the occurrence of false intersections or ghosts. However, this problem occurs only if the sensor system and the targets are on a common plane. Thus, one approach is to define a common reference plane and to use the angles with respect to this plane as discriminants to eliminate ghosts. Reference [10] gives other techniques for reducing the problem of ghosts.

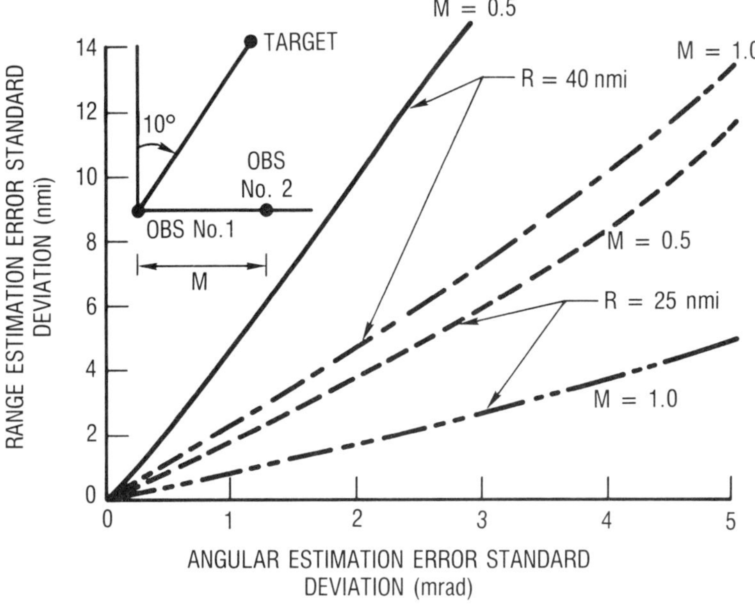

Figure 7.9 Triangulation range estimation accuracy.

7.7 SENSOR ALLOCATION

The simplest approach to sensor allocation uses one sensor to cue another. For a typical scenario, a low angular resolution sensor with a large field of view obtains a preliminary estimate of target position. Then, a second, high resolution (angular resolution) sensor is slewed to the predicted target position [16]. This approach may be appropriate if the number of tracks requiring sensor allocation is limited. However, a dense target environment requires more sophisticated approaches.

A modern sensor allocation algorithm will be required to adaptively direct multisensor systems to (1) search for new targets, (2) update target positions, (3) identify targets, and (4) remain covert and thus evade detection by the enemy. This algorithm should be adaptive to varying target densities and dynamics, to sensor capabilities, and to the overall situation and mission objectives.

There are two general approaches to sensor allocation in a dense target environment. The first is based upon longer-term planning and involves computing the expected frequency at which the many required

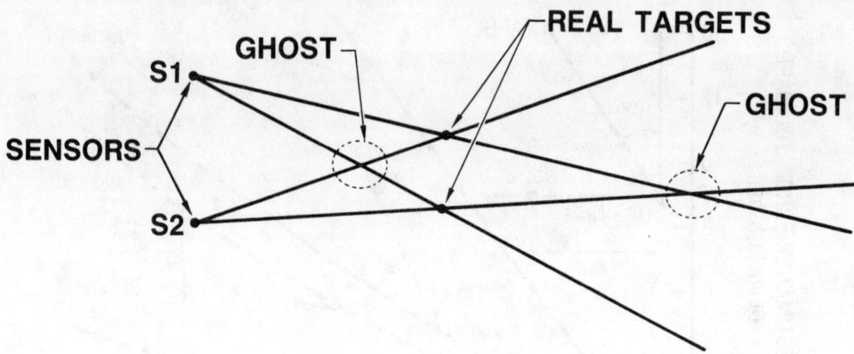

Figure 7.10 Ghost track formation with sensors and targets on a common plane.

tasks should be performed. For example, a regular update rate would be computed for each target track. The next step will involve bin or time slot packing to accommodate long-term objectives performed by each sensor.

An alternative approach is to allocate more dynamically based upon current conditions. For example, rather than adhere to some predetermined track update rate, the desirability of updating each track is computed regularly based upon current conditions and compared with the desirability of performing other tasks, such as search or communication. Then, short-term objectives are formed and updated frequently. We will next discuss two techniques available for implementing the dynamic sensor allocation approach.

7.7.1 Utility Theory Based Allocation

Sensor allocation can be based upon the minimization of some cost function or, equivalently, the maximization of a utility function. The basic approach here is to determine the usefulness of the state of knowledge before and after the potential application of a sensor to a given task. For example, considering the track update allocation option, the usefulness, or the utility, of the estimate of track position can be measured by the Kalman covariance matrix. Then, the utility of the track update can be computed from the present covariance matrix and the computed covariance matrix if an observation is received. Similarly, the utility of search can be determined from the expected value of the targets to be detected in search. Reference [2] discusses the utility theory approach in more detail and includes several examples.

Subsequent experience has shown this approach to be particularly effective when used with target maneuver detection logic. Upon maneuver detection, a standard approach is to increase the elements of Kalman filter covariance matrix, forcing more frequent track updates. Finally, as discussed in Reference [17], the overall problem of allocating multiple sensors to multiple tasks based upon cost or utility function can be considered as an application of the well-known transportation problem.

7.7.2 Sensor Allocation Based on Expert Systems

The use of expert system techniques for sensor allocation (References [18] and [19]) offers several important advantages. First, a rule-based system readily allows for expansion through the inclusion of additional rules. The use of rules also makes it easier to interpret and explain the behavior of an expert system, as opposed to analyzing the properties of utility functions. Finally, the expert system technique conveniently allows for sensor allocation to be closely related to situation assessment.

We will use a brief summary of the contents of Reference [19] as an example of the application of expert system methods to the sensor allocation problem. The method starts by defining quantitative (root) concepts such as distance, closing rate, and target bearing, which are determined from encounter geometry. Following this, qualitative (higher-order) concepts, such as target *identification* (ID) and danger posed by the target, are defined. The membership of each target in the "fuzzy set" defined by these concepts is computed. For example, a simple relationship of time-to-go ($R/\dot{R}$) can be used for the root concept "close." Then using a set of "IF-THEN" rules the membership of each target in the fuzzy set defined by higher order concepts, such as degree of danger, is computed. Finally, using a combination of root and higher-order concepts, the membership, as well as priorities, in the set of tracks to be updated is computed and allocation of sensors is made.

The allocation logic for sensors in the expert system is closely related to situation assessment, since the examined concepts, such as position, proximity of the target, and hostile intent, are also integral components of situation assessment.

7.8 OTHER ISSUES

This section briefly discusses some of the other important issues involved in the development of systems that utilize multiple sensors. The problem of sensor alignment [20] is probably the least understood and documented aspect of multiple sensor systems. The question of determining when the declaration of target ID can be made is discussed in [21] and [22]. Reference [23] discusses how multiple sensor performance requirements can be determined, based on overall system requirements. In the next chapter, we discuss the evaluation of multiple sensor systems for tactical operations. Finally, Reference [24], in addition to an overview of multiple sensor fusion, discusses computational issues.

This chapter has been concerned primarily with the process of track formation and the estimation of individual target characteristics once tracks are formed. However, much recent interest has centered on the development of *artificial intelligence* (AI) techniques that use the track files along with other data bases to form higher-level inferences. These applications include determining interrelationships among targets, implying target intent, and in general, predicting longer-term enemy target intent and behavior. We will briefly discuss how this can be done using the flow chart of Figure 7.11, which was derived for an avionics system. More detail of similar architecture is given in Reference [25].

In Figure 7.11, the *multiple sensor fusion* (MSF) function program processes the input sensor data to form track files using the methods discussed in this chapter. Note that, in general, this may include fusion of target tracks, as well as measurements. The first AI function, called *object analysis*, uses the track files to infer additional information about single objects (targets) or small groups of targets. The block diagram continues with target track files and situation assessment, which constitutes the second AI function. Situation assessment draws more global inferences about the behavior of all the tracks, based upon the track file information and the output of object analysis. The information from situation assessment is used for sensor allocation and may, in general, be used for other higher-order decisions such as tactics (should a maneuver be performed to evade hostile threats *et cetera*). Finally, sensor allocation (as discussed previously) determines the manner in which sensor resources are used to collect further data. Note that there may also be other allocation functions, such as for computational resources or *electronic countermeasures* (ECM). Finally, nearly all systems will allow for operator (in this case the pilot) input to sensor allocation or, potentially, to situation assessment as well.

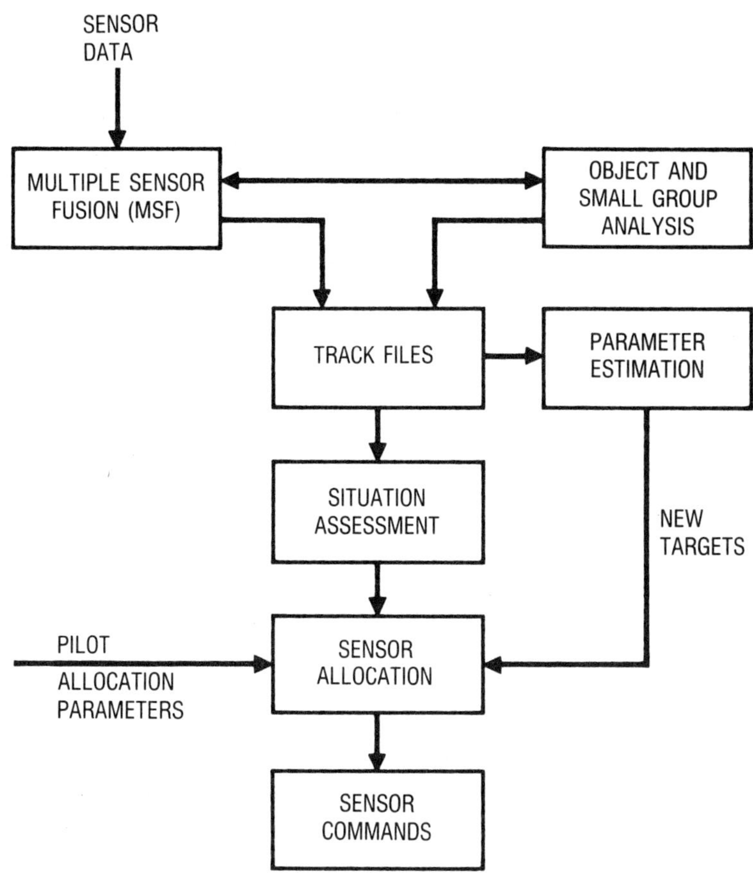

Figure 7.11 Multiple sensor, multiple target system flow chart.

7.9 SUMMARY

This chapter has outlined the important issues and the methods applicable to the association and fusion of multiple sensor data. In the past, simple data association methods, such as that of the nearest neighbor, have been used for track formation. In the future, increased computational capabilities should lead to much wider application of multiple hypothesis tracking methods.

Present multiple sensor systems tend to use sensor level tracking with each sensor effectively operating autonomously and displaying its track information for an operator to intercept. Future systems will also seek autonomous sensors to the extent that a backup is provided in the case of failure with the computer or other sensors. However, the advantage of the central level tracking approach will mean that more systems will combine both sensor and central level tracking, as discussed in Reference [26].

Except for simple tracking conditions, such as air traffic control, the use of Kalman filtering has essentially replaced fixed gain filtering methods of α-β-γ types. Kalman filtering will be particularly necessary for multiple sensor systems. Bayesian or Dempster-Shafer methods should soon replace more heuristic methods, such as possibility theory, for combining attribute data for target ID analysis.

More sophisticated sensor allocation methods will be required for future systems. Also, there will be an increasing use of AI techniques for interpreting the track file information. Thus, it is likely that expert systems approaches will be applied widely.

REFERENCES

1. A. Farina and F. A. Stader, *Radar Data Processing*, Vols. 1 and 2, John Wiley, New York, 1985.
2. S. S. Blackman, *Multiple Target Tracking with Radar Applications*, Artech House, Norwood, MA, 1986.
3. T. E. Fortmann, Y. Bar-Shalom, and M. Scheffe, "Sonar Tracking of Multiple Targets Using Joint Probabilistic Data Association," *IEEE J. Ocean Engr.*, July 1983, pp. 173–184.
4. P. Smith and G. Buechler, "A Branching Algorithm for Discriminating and Tracking Multiple Objects," *IEEE Trans. Automatic Control*, February 1975, pp. 101–104.
5. D. B. Reid, "An Algorithm for Tracking Multiple Targets," *IEEE Trans. Automatic Control*, December 1979, pp. 843–854.
6. G. V. Trunk and J. D. Wilson, "Association of DF Bearing Measurements with Radar Tracks," *IEEE Trans. AES*, July 1987, pp. 438–447.
7. Y. Bar-Shalom and L. Campo, "The Effect of Common Process Noise on the Two Sensor Fused-Track Covariance," *IEEE Trans. AES*, November 1986, pp. 803–805.
8. D. L. Alpach, "An Approach to Multi-Sensor Integration Problem," *Proc. EASCON*, September 1983, pp. 149–152.
9. N. C. Moharty, *Random Signals Estimation and Identification*, Van Nostrand Reinhold, New York, 1986.

10. G. Van Keuk, "Deghosting in an Automatic Triangulation System," *Proc. Int. Radar Conf.*, October 1982, pp. 291–295.
11. D. V. Stallard, "An Angle-Only Tracking Filter in Modified Spherical Coordinates," *Proc. 1987 AIAA Guidance, Navigation and Control Conf.*, August 17–19, 1987, pp. 542–550.
12. A. P. Dempster, "A Generalization of Bayesian Inference," *J. Royal Statistical Soc.*, Volume 30, 1968, pp. 205–247.
13. G. Shafer, *A Mathematical Theory of Evidence*, Princeton University Press, Princeton, NJ, 1976.
14. B. H. Cantrell and A. Grindley, "Multiple Site Radar Tracking System," *Proc. IEEE Int. Radar Conf.*, April 1980, pp. 348–354.
15. R. T. Lacoss, "Distributed Mixed Sensor Aircraft Tracking," *Proc. American Control Conf.*, June 1987, pp. 1827–1830.
16. T. G. Haskins, "Sensor Cueing Performance Analysis," *Proc. NAECON*, May 1984, pp. 262–265.
17. J. M. Nash, "Optimal Allocation of Tracking Resources," *Proc. IEEE Conf. Decision and Control*, December 1977, pp. 1177–1180.
18. E. R. Carney and J. J. Bourquin, "An Expert System for Sensor Allocation in Multi-sensor Integration," *Proc. American Control Conf.*, June 1987, pp. 928–933.
19. R. Popoli and S. S. Blackman, "Expert System Allocation for the Electronically Scanned Radar," *Proc. American Control Conf.*, June 1987, pp. 1821–1826.
20. W. G. Bath, "Association of Multiple Radar Data in the Presence of Large Navigation and Sensor Alignment Errors," *Proc. IEEE Int. Radar Conf.* (London), October 1982, pp. 169–173.
21. P. J. Nahin and J. L. Pokoski, "NCTR plus Sensor Fusion Equals IFFN or Can Two Plus Two Equal Five," *IEEE Trans. AES*, May 1980, pp. 320–337.
22. S. S. Blackman, T. A. DuPuis, and A. J. Mendez, "Application of Multi-Sensor Correlation Techniques to the Target Validation Problem," *Proc. First DoD Tri-Service Conf. Total Target Indentification Tech.* (Fort Monmouth, NJ), October 1981, pp. 699–713.
23. M. P. Dana, "Establishment of Air Defense Sensor Requirements for Automatic Aircraft Tracking," *AGARD Conf. Proc. No. 252, Strategies for Automatic Track Initiation*, Monterey, CA, October 1978, pp. 19-1–19-20.
24. E. L. Waltz, "Computational Considerations for Fusion in Target Indentification Systems," *Proc. NAECON*, May 1981, pp. 492–497.
25. E. L. Waltz and D. M. Buede, "Data Fusion and Decision Support for Command and Control," *IEEE Trans. Systems, Man and Cybernetics*, November–December 1986, pp. 865–879.

26. S. C. A. Thomopoulos, R. Viswanathan, and D. C. Bougoulias, "Optimal Decision Fusion in Multiple Sensor Systems," *IEEE Trans. AES*, September 1987, pp. 644–653.

Chapter 8
Multiple-Sensor Effectiveness and the Expert Matrix Method

Sensor suites may consist of a single sensor, such as a thermal imager or a microwave radar, or a set of sensors, such as a thermal imager and a laser radar ranging unit. As discussed in Chapter 7, multiple sensor systems are usually composed of complementary sensors. For example, in the thermal imager and laser radar combination, the former is used to detect targets in the passive mode and the latter is used to obtain target range. In a number of cases multiple sensors are used for target identification. Reference [1] cites the case where detection and identification of armored targets are perfromed by three complementary sensors. These sensors are a laser ranging radar that measures the size of the target on the ground, an infrared sensor that determines the presence of a hot spot indicating an engine, and a receiver at MMW frequency to determine whether the object is metal, which reflects solar-created energy. If all three criteria are met, the object is classified as an armored target.

The effectiveness of sensor systems against specified requirements are usually studied on a qualitative basis. Although several semianalytic and computer simulation methods are available for these studies, none of these methods results in a quantitative ranking of alternative sensor suites in a given application. In this chapter we use the expert matrix method to obtain a quantitative ranking of examined sensor systems. This method uses analytical and simulation results, as well as expert inputs.

8.1 INTRODUCTION

The quantitative evaluation of sensor effectiveness has presented a formidable task to operation analysts, mission planners, and sensor designers. Multiple sensor systems compound the problem many fold, as various sensors do not provide the same type of target or scene information. Methods should be devised for quantitative evaluation of sensor systems against specified requirements. The solution of this problem will involve a study of the performance of individual sensors, as well as combinations of sensors. As shown in Figure 8.1, these sensors may include thermal imager systems (TIS), millimeter wave (MMW) radars, infrared search and track sets (IRST), laser radar systems (LRS), and radar warning receivers (RWR). Radar warning receivers are devices which detect radar transmission, alerting the user of the presence of another radar system in the area.

As shown in Figure 8.1, the performance evaluation methods of individual sensor systems should be followed by performance evaluation of combination of sensors. These may include MMW plus TIS or TIS plus IRST and so on. Having obtained the performance models of individual and combinations of sensors we can proceed to examine the applicable methodology for their effectiveness against given requirements. The existing methods of effectiveness evaluation include scenario methods, engagement models, and the Bayesian approach. Future methodology may include quantitative application of methods used in AI and expert systems. These methods are a combination of analytical studies and expert inputs. The quantitative effectiveness analysis should also include specific requirements of cost, weight, and volume, as shown in the fifth block of Figure 8.1.

At the conclusion of a quantitative effectiveness study, the results should indicate the level of conformity of the sensor suite to the requirements. Typical results of a study of five sensor suites may be expressed as in Table 8.1. This table shows the comparison of five sensor suites, A through E, against specified requirements. The relative performance column indicates the performance of each sensor suite against the specified requirements. The values of the table were normalized with respect to suite A. The relative cost column is a weighted measure of *cost*, *weight*, and *volume* considerations. Using this table, we can make meaningful comparisons. For example, sensor suite D will be 23 percent more effective than sensor suite A against specified requirements, but it will "cost" 40 percent more.

In this chapter we will outline ideas and concepts toward solving the multiple sensor effectiveness problem. It is realized that the complete solution of this problem is a long way away. However, the methods of this chapter can be considered a first step in the direction of quantitative effectiveness analysis. In order to confine the problems, the discussed sensor systems will be limited to short-range sensors, such as those used in helicopters and manned tactical systems.

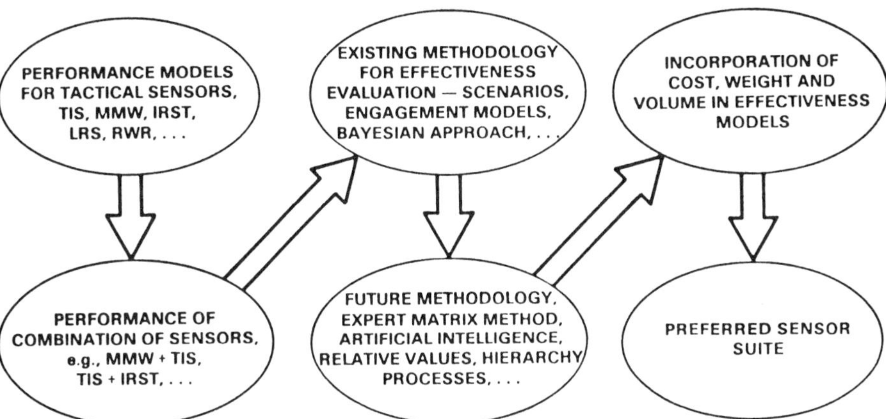

Figure 8.1 Elements of multiple sensor effectiveness study.

Table 8.1 Example of Results of Typical Multiple Sensor Effectiveness Study

Sensor Suite	*Relative Performance*	*Relative Cost*
A	1.0	1.0
B	0.84	0.67
C	0.30	0.30
D	1.23	1.40
E	0.61	0.23

8.2 SENSORS

A broad description of the characteristics of these sensors and their approximate operating ranges is given in Figure 8.2. All of the sensors

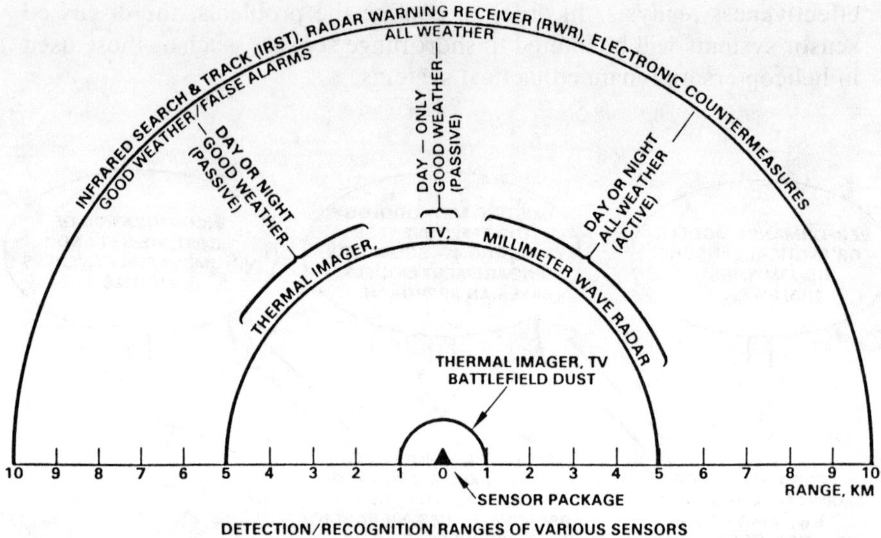

Figure 8.2 Individual sensors forming sensor suites.

fall into the two broad categories: passive and active. Some sensors, such as the TV-like ones, depend on outside illumination for their operation and hence can operate only in the daytime. Others, such as thermal imagers, as described in Chapters 1 and 5, operate on thermal gradients during both day and night. Thermal imager and TV-like sensors, because of their short operating wavelengths (in the μm) produce high resolution images. Because of the short wavelengths, however, these sensors become degraded under battlefield dust, smoke, and bad weather. Radar sensors, such as MMW radars with longer wavelengths (in the mm region), can operate satisfactorily despite dust, smoke, or bad weather, but they do not possess the high resolution capability of thermal imagers and TV-like sensors. Additionally, radar sensors transmit their own energy and are active devices, a concern that should be addressed in countermeasure studies.

Considering all of these factors and the countermeasures against these sensors, it can be readily concluded that the selection of sensors to fulfill the requirements of a given mission is a very crucial part of the design of any surveillance, target acquisition, and fire control system.

8.3 EFFECTIVENESS EVALUATION, EXISTING METHODOLOGY

Existing methods of effectiveness evaluation include scenario evaluation method, engagement computer models, and Bayesian analysis. The scenario evaluation method is one of the original methods applied to the effectiveness evaluation of a given sensor system in a battlefield situation. The method begins by laying out a scenario resembling, as closely as possible, the battle situation where the sensors under consideration will be used. Sensor performance characteristics (such as probability of target detection *versus* range) obtained from the performance models are used.

Starting from an initial encounter situation and using these sensor performance characteristics, a sequence of decisions is made as a function of time by the opposing forces, much like in any game situation. The quantitiative effectiveness of the deployed sensor system is evaluated by the outcome of the battle. The same evaluation procedure can be repeated using another set of sensors with applicable performance characteristics. Comparison of these results will determine the effectiveness of various sensor systems.

The engagement model method consists of a Monte Carlo digital computer program that simulates defensive, offensive, and meeting engagements between opposing forces. The program uses sensor performance values generated by mathematical performance models of sensors or combinations of sensors. The output of the engagement model gives the quantitative effectiveness of various sensor suites under specific battle conditions. These results can be used for ranking the effectiveness of examined sensor systems, as well as general background material for the expert matrix method.

Another multiple sensor effectiveness evaluation approach is that of the Bayesian analysis, as discussed in Chapter 7 and Reference [2]. In this method, we describe the probabilistic behavior of various sensor systems and mathematically evaluate the probabilistic behavior of combinations of these sensors. These methods are usually applied in cases where several sensors of the same kind are netted. For example, consider a radar net consisting of three identical radars located on a triangular grid. Having the probability of identification of targets by each radar, the Bayesian approach will give the confidence level of correct target identification when one, two, or all three of the radars identify the target.

Although the existing methods of sensors effectiveness evaluation are extremely useful in understanding the workings of a sensor system in a given scenario, they fail to provide a quantitative measure of sensor effectiveness. By using artificial intelligence methods and ideas, another procedure for quantitative evaluation of sensor systems is developed, called the *expert matrix method*. Using this method and drawing upon experience gained from the scenario method, engagement models, and expert opinions, we can quantitatively describe the sensor suite effectiveness, as discussed in the next section.

8.4 EXPERT MATRIX METHOD

An expert is a system AI program that relies on a body of knowledge to perform a difficult task that can usually be performed only by human experts (Richard Forsyth, Editor, *Expert Systems*, Chapman and Hall, 1984). In this respect, it is different from the analytical methods, where classical mathematics is used to relate a sequence of events as a function of distance and time. Exercise of analytical procedures, however, contributes to expert system analysis in providing the vital knowledge base.

In the expert matrix method of this section, this human thought process is formalized and expressed analytically. After a series of individual evaluations of parameters affecting the final decision, the entire decision process is expressed mathematically in a matrix format. Computerized matrix operations result in a quantitative ranking of alternate sensor suites.

The expert matrix method just discussed briefly is applied to the multisensor effectiveness evaluation to obtain quantitative ranking for the individual and combination sensors used in a given scenario. The method separates the performance of sensors from battle requirements of a given scenario. Using information gathered from the previously described scenario evaluation method, the engagement model method, and expert inputs, we can develop a sensor-condition matrix. This matrix will describe the effectiveness of the sensor performance under realistic battlefield conditions. Following this, a condition scenario is developed based on the scenario requirements. Multiplication of the sensor-condition matrix by the condition-scenario matrix will give, in effect, the correlation between the sensor performance and the scenario requirements. This product will rank the sensors as to their effectiveness with respect to the given scenario.

8.4.1 Theoretical Background

The dependence of functions can be expressed in terms of correlation integral as follows:

$$\varphi_{fg} = \int_0^X f(x)g(x)\mathrm{d}x \tag{8.1}$$

where $f(x)$ and $g(x)$ are functions whose interdependence is to be evaluated and φ_{fg} is the correlation function. Figure 8.3 shows continuous and discrete representations of functions $f(x)$ and $g(x)$. The correlation integral can be represented in discrete form by the summation

$$\varphi_{fg} = \sum_{i=1}^{N} f_i g_i \Delta x \tag{8.2}$$

Thus, from Eq. (8.2) the sum of the products of discrete values of f_i and g_i will be proportional to the correlation function. For identical functions, that is, $g(x) = f(x)$, Eq. (8.2) results in

$$\varphi_{fg} = f_1^2 + f_2^2 + \cdots + f_N^2 \tag{8.3}$$

which is the sum of the square of the values of the function $f(x)$ at intervals Δx. In summary, the correlation function presents an indication of similarity of two functions, $f(x)$ and $g(x)$. Higher values of the correlation function indicate a greater degree of similarity. The discrete summation of Eq. (8.2) can be considered as the elements of product matrices. Consider the product matrices

$$\underset{(m \times 1)}{C} = \underset{(m \times n)}{S} \; \underset{(n \times 1)}{R} \tag{8.4}$$

where S is a matrix, rows of which represent discrete values of functions $f_1, f_2, \ldots$; R is a column matrix representing $g_1, g_2, \ldots$; and C is the correlation matrix. The elements of the C matrix can be represented as

$$c_{i1} = \sum_{j=i}^{N} s_{ij} r_{j1} \tag{8.5}$$

where the s_{ij} denotes discrete values of f_j and the r_{j1} denotes discrete values of g_1.

Using matrix notation, Eq. (8.4), the entire correlation process between several alternatives (sensor suites) represented by rows of matrix S and a set of requirements represented by the column matrix R can be put in a matrix product form as follows:

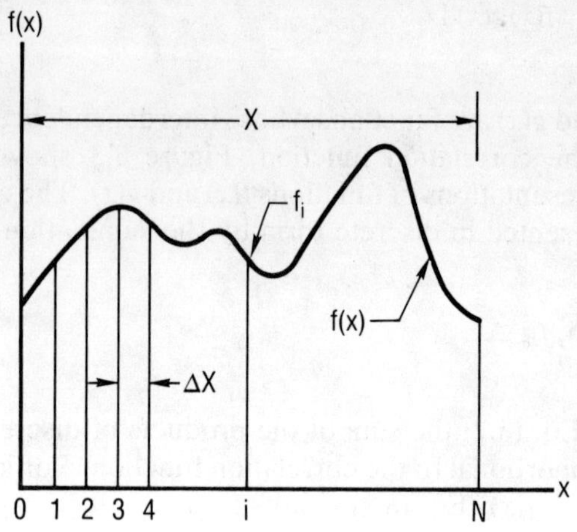

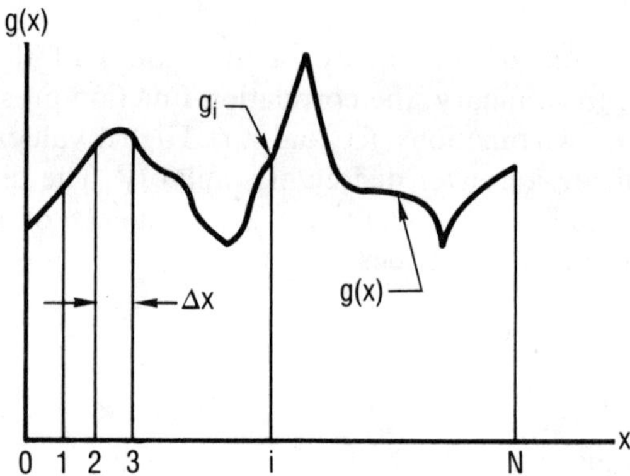

Figure 8.3 Representation of functions $f(x)$ and $g(x)$.

$$\begin{pmatrix} s_{11} & s_{12} & s_{13} & \cdots & s_{1n} \\ s_{21} & s_{22} & s_{23} & \cdots & s_{2n} \\ & & \cdots & & \\ s_{m1} & s_{m2} & s_{m3} & \cdots & s_{mn} \end{pmatrix} \begin{pmatrix} r_{11} \\ r_{21} \\ \cdots \\ r_{n1} \end{pmatrix} = \begin{pmatrix} c_{11} \\ c_{21} \\ \cdots \\ c_{m1} \end{pmatrix} \tag{8.6}$$

We will see in the following discussion that, in sensor effectiveness analysis, the rows of matrix S represent sensor suite characteristics, the column matrix R represents the requirements, and the product matrix C

represents the degree of correlation of sensors and requirements. The most effective sensor system conforming to the requirements of column matrix R of (8.6) will have a correlation value of

$$r_{11} + r_{12}^2 + r_{13}^2 + \ldots + r_{n1}^2 = N \tag{8.7}$$

where N is the normalization constant. The correlation matrix C can be normalized with respect to the numerical value of N of Eq. (8.7). The following numerical example will illustrate this procedure.

8.4.2 Numerical Example

In an application of the expert matrix method, we form a sensor-condition matrix describing the performance of the sensors under specified battlefield conditions. Figure 8.4 lists several types of sensors and a number of operating conditions and parameters for these sensors. The sensors include thermal *imaging systems* (TIS), MMW radars, IRSTs, and combinations of sensors, such as thermal imagers and MMW radars. The operating conditions and parameters may include, but are not limited to, smoke, dust, weather effects, search volumes, and so on. The weight volume and cost parameters are also given at the end of conditions parameters. These parameters will be considered separately in the following formulation.

As a specific numerical example, consider the following five sensors or combination of sensors.

Thermal imager system (TIS)
Millimeter wave radar system (MMW)
Infrared search and track set (IRST)
Thermal imager and millimeter wave radar combination (TIS) and MMW)
Radar warning receiver (RWR)

It is assumed that these sensor systems are required to operate under the following battlefield conditions—smoke and dust, adverse weather—as well as to satisfy the following encounter requirements:

Large search volumes
Target recognition
Target range value
Target angular position
Countermeasures

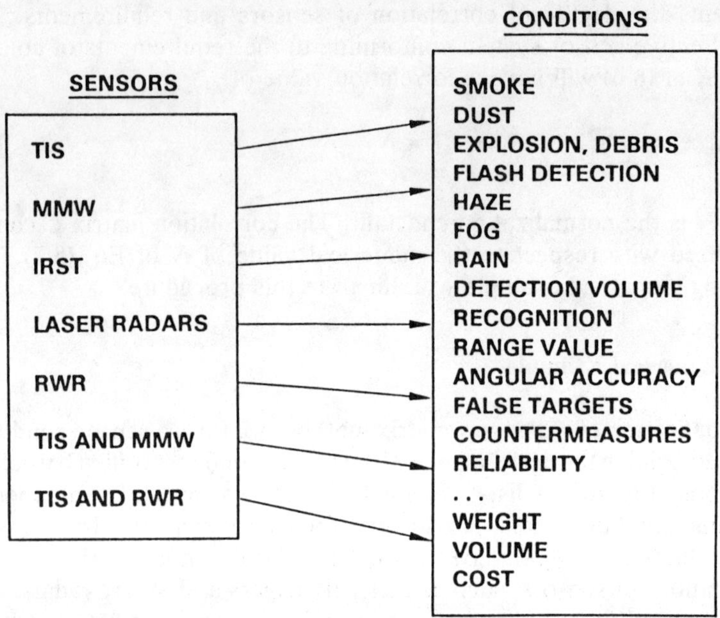

Figure 8.4 Sensors and operating conditions and parameters.

Table 8.2 gives the *sensor-condition matrix* with rows 1 through 5 containing the sensors and columns 1 through 7 describing battlefield conditions and encounter requirements as given earlier. The performance of the sensor under the given condition is graded from 1 to 10, with 10 being the best. Expert system knowledge is used to determine the values of the matrix of the table. For example, considering that thermal imagers do not perform well under smoke, dust, and adverse weather conditions, values of 3 and 5 are assigned, respectively, to the corresponding elements of the matrix. On the other hand, thermal imagers can recognize targets and obtain very accurate angular position data; therefore, values of 10 and 10, respectively, are assumed. Similarly, MMW radars are capable of large volume search and operate in adverse weather, and they are *not* severely affected by smoke or battlefield dust; therefore, scores of 9 are used for the corresponding elements of the matrix.

The scores for a combination of sensors, such as TIS plus MMW, are determined by taking the larger of the individual scores for each category. We realize that this step may not be representative of the actual performance of the combination of sensors, but it is considered a good first step. Thus, from a knowledge of sensors, the sensor-condition matrix of Table 8.2 can be completed.

Table 8.2 Sensor-Condition Matrix

Sensor	*Smoke, Dust*	*Weather*	*Search Volume*	*Recognition*	*Range Value*	*Angular Value*	*Countermeasures/ False Alarm*
TIS	3	5	3	10	5	10	10
MMW	9	9	9	1	9	5	5
IRST	3	3	9	1	1	1	3
TIS and MMW	9	9	9	10	9	10	10
RWR	9	9	3	1	1	1	10

The importance of operating conditions to a given scenario can also be quantitatively described, as shown in the column matices of Table 8.3. In this example, two scenarios were selected and the importance of each condition to the given scenario was scored on the basis of 10 points. It was assumed that scenario 1 occurs during winter and scenario 2 during summer. Thus, the importance of operation under battlefield dust in summer is underscored by a value of 10, while in the winter, because of wetness and reduced amount of dust, a value of 5 is used. Otherwise, the two scenario requirements are similar.

Table 8.3 Condition-Scenario Matrix

Condition	*Scenario* *1* (Winter)	*Scenario* *2* (Summer)
Smoke/dust	5	10
Weather	3	5
Search detection	5	5
Recognition	10	10
Range value	3	3
Angular value	10	10
Countermeasures/false alarms	10	10

The relative sensor effectiveness values can be obtained by multiplying the *sensor-condition matrix* of Table 8.2 by the *condition-scenario column matrices* of Table 8.3. Note that the sensor-condition matrix is a 5 × 7 matrix and the scenario-condition matrix, consisting of a single column, is a 7 × 1 matrix. The resulting product matrix will be a 5 × 1 matrix describing the rank of the sensor system. This product matrix quantitatively represents the correlation between the sensor capabilities and scenario requirements. For each scenario, a 5 × 1 matrix will be obtained as shown in Table 8.4. Note that values of this table are normalized with respect to the thermal imager systems.

For example, the relative values of Table 8.4 show that the TIS plus MMW is 23 percent more effective in the case of scenario 1, under winter battlefield conditions, and 32 percent more effective in the case of scenario 2, under summer battlefield conditions, than a thermal imager system alone. This situation is to be expected, since one of the main attributes of MMW radar is its accuracy despite battlefield dust and smoke, which is greater in summer. We see further that IRST and RWR sensors alone are not very effective.

Table 8.4 Sensor Performance Rankings Obtained from Matrix Multiplication

Sensors	*Scenario 1* (Winter)	*Scenario 2* (Summer)
TIS	1.00	1.00
MMW	0.70	0.82
IRST	0.34	0.37
TIS and MMW	1.23	1.32
RWR	0.58	0.70

At this point, the question may arise, What do we mean when we say system A is 20 percent more effective than system B? The answer to this question is very complicated and implicit in the values assumed for the sensor-condition matrix of Table 8.2 and the condition-scenario matrix of Table 8.3. Thus, the degree of effectiveness is a quantitative representation of the assumptions made for the combination of sensor performance measures and scenario requirements.

The sensor rankings of Table 8.4 were normalized with respect to the TIS, as discussed earlier. Another normalization constant connected to scenario requirements may also be used. Consider scenario 1 of Table 8.3. The optimal sensor suite for these requirements will be tailor-made for this scenario, with parameters matched to scenario requirements and a normalization constant consisting of the squares of the column matrix:

$$(5)^2 + (3)^2 + (5)^2 + \ldots + (10)^2 = 368 \tag{8.8}$$

This normalization constant will relate the degree of conformity of sensors to those of the scenarios.

8.4.3 Weight, Volume, and Cost

Weight, volume, and cost can also be incorporated in effectiveness studies by weighted matrix considerations. The method is to reduce these three factors to a single number for a given sensor system. Consider the five sensor suites of the previous example with the following estimates: weight (in lbs), volume (in cubic ft), and cost (in thousands of dollars)

	Weight	Volume	Cost
TIS	150	2	250
MMW	100	1.5	150
IRST	50	0.75	50
TIS AND MMW	220	3	320
RWR	50	0.5	40

(8.9)

Normalizing the matrix of (8.9) to the parameters of TIS system, we have

	Weight	Volume	Cost
TIS	1	1	1
MMW	0.66	0.75	0.6
IRST	0.33	0.375	0.2
TIS and MMW	1.46	1.5	1.28
RWR	0.33	0.25	0.16

(8.10)

At this point the importance of each parameter (weight, volume, and cost) should be decided against the particular requirements of the sensor system. For example, in a ground-based system increased weight may not be as important as increased volume and cost. Thus, assuming a weighting matrix of

$$\begin{bmatrix} 0.50 \\ 1.00 \\ 1.00 \end{bmatrix} \tag{8.11}$$

for weight, volume, and cost, we can calculate the system "cost" matrix by matrix multiplication of Eqs. (8.10) and (8.11). This results:

$$\begin{bmatrix} 2.50 \\ 1.68 \\ 0.74 \\ 3.51 \\ 0.51 \end{bmatrix} \tag{8.12}$$

Normalizing the weighted cost matrix of (8.12) by using 2.5 as the normalization constant, we have

$$\begin{array}{l} \text{TIS} \\ \text{MMW} \\ \text{IRST} \\ \text{TIS and MMW} \\ \text{RWR} \end{array} \begin{bmatrix} 1.00 \\ 0.67 \\ 0.30 \\ 1.40 \\ 0.23 \end{bmatrix} \tag{8.13}$$

With the values of Eq. (8.13), we now have both sensor effectiveness measures in performance and "cost" effectiveness values. Using the performance measures of scenario 1 from Table 8.4 with the cost values of Eq. (8.13), we obtain Table 8.5. The values of this table specify the quantitative effectiveness measures of the five sensor suites in this specific application. The table shows that TIS plus MMW will, for example, have a 23 percent better performance than the TIS system alone, but this additional performance will increase the weighted "cost" of the system by 40 percent.

Table 8.5 Performance and Weighted Cost Measures of the Example Problem

System	*Performance*	*Weighted Cost*
TIS	1.0	1.0
MMW	0.70	0.67
IRST	0.34	0.30
TIS plus MMW	1.23	1.40
RWR	0.58	0.23

The quantitative ranking of the expert matrix method may be further improved by the relative weights method of T. L. Saaty (References [3] and [4]) as we will describe.

8.5 RELATIVE WEIGHTS METHOD

The relative sensor ranking obtained by the expert matrix method of the last section can be improved by the relative weight method of this section. The relative weight method is designed to solve a fundamental problem in decision theory: how to derive weights for a set of activities according to their importance. The idea is to initially estimate the matrix of relative weights and then obtain a more accurate estimate by calculating the eigenvector (characteristic vector) of this matrix. We can see that, if the initial assumed relative weight vector is absolutely correct, then the calculated eigenvector (characteristic vector) will be identical to it. On the other hand, errors and inconsistencies in the original assumption will be reduced and distributed in the calculated eigenvector representing the relative rankings. Using the relative weight method, the sensor ranks of the expert matrix method can be used as an initial approximation. This approximate rank is then improved by the relative weight matrix calculations.

8.5.1 Theoretical Background

Consider the relative weight matrix of a set of four sensor systems S_1, S_2, S_3 and S_4, as shown in Table 8.6.

The 4 × 4 relative weight matrix in this table represents a consistent set of relative weights, with the property

$$a_{ij} = 1/a_{ji} \tag{8.14}$$

where i and j are row and column numbers and a denotes the elements of the matrix. The diagonal matrix elements are unity as relative weight of each sensor with respect to itself is 1. The eigenvector of the matrix in Table 8.6 will represent the relative weights of the sensors. The eigenvector can be calculated to be

$$\begin{pmatrix} 1 \\ W_2 \\ W_3 \\ W_4 \end{pmatrix} \tag{8.15}$$

Table 8.6 Relative Weight Matrix for Example Problem

Sensors	S_1	S_2	S_3	S_4
S_1	1	$1/W_2$	$1/W_3$	$1/W_4$
S_2	W_2	1	W_2/W_3	W_2/W_4
S_3	W_3	W_3/W_2	1	W_3/W_4
S_4	W_4	W_4/W_2	W_4/W_3	1

This calculation can be verified using the definition of eigenvectors as in Table 8.7, where "4" represents the eigenvalue. Thus, the relative weight matrix of Eq. (8.15) is identical to the eigenvector of the relative weight matrix. Since the relative weight matrix is consistent (that is, for example, the relative weight of S_4 to S_2 is W_2/W_4 and S_3 to S_2 is W_2/W_3, then S_4 to S_3 is W_3/W_4), the eigenvalue will be equal to the rank of the relative weight matrix of Table 8.6. Saaty argues in References [3] and [4] that inconsistencies in the values of the components of Table 8.6 will make the eigenvalue different from the number representing the rank of the matrix.

Table 8.7 Definition of Eigenvectors in the Example Problem

Relative Weight Matrix	*Relative Weights*		*Eigenvector*
$\begin{pmatrix} 1 & 1/W_2 & 1/W_3 & 1/W_4 \\ W_2 & 1 & W_2/W_3 & W_2/W_4 \\ W_3 & W_3/W_2 & 1 & W_3/W_4 \\ W_4 & W_4/W_2 & W_4/W_3 & 1 \end{pmatrix}$	$\begin{pmatrix} 1 \\ W_2 \\ W_3 \\ W_4 \end{pmatrix}$	$= 4$	$\begin{pmatrix} 1 \\ W_2 \\ W_3 \\ W_4 \end{pmatrix}$

8.5.2 Example: The Wealth of Nations

The relative weight method developed by T. L. Saaty has been applied to rank nations as to their gross national products. Historically, this was one of the first examples illustrating the power of this method. It will be shown that, by using the approximate relative values describing the wealth of one nation with respect to another in a matrix format and computing the characteristic vector of this matrix, a reasonable approximation to the relative economic ranking of these nations can be obtained.

Consider the following countries:

United States (US)
Japan
Federal Republic of Germany (FRG)
France
United Kingdom (UK)
Italy
Canada

Assume the economic relative rankings given in Table 8.8 for these countries. The ranking matrix of the table was obtained primarily by

Table 8.8 Ranking of Nations in Example

	US	*Japan*	FRG	*France*	UK	*Italy*	*Canada*
US	1	1/3	1/5	1/7	1/8	1/9	1/10
Japan	3	1	1/2	1/3	1/3	1/4	1/5
FRG	5	2	1	3/4	3/4	1/2	1/2
France	7	3	4/3	1	7/8	1/2	1/2
UK	8	3	4/3	8/7	1	3/4	3/4
Italy	9	4	2	2	4/3	1	1
Canada	10	5	2	2	4/3	1	1

considering the population of these countries and their relative industrial bases. As an example, France is considered to be one-seventh as strong economically as the US and one-third as strong as Japan. There is a certain degree of inconsistency in the matrix of Table 8.8. For example, if the US is three times as strong as Japan and France is one-seventh as strong as the US, then France will be three-sevenths as strong as Japan. However, the matrix of the table specifies France as one-third as strong as Japan. It is argued [3,4] that these inconsistencies will work out in the final solution, and at this stage, it is more imortant to incorporate common notions than precise data.

Table 8.9 Characteristic Vector and GNP of Several Countries

Countries	*Characteristic Vector*	GNP *Ratio 1985*
US	1.0	1.0
Japan	0.34	0.40
FRG	0.17	0.19
France	0.13	0.15
UK	0.12	0.13
Italy	0.094	0.095
Canada	0.088	0.090

The characteristic vector (eigenvector) of this matrix is calculated using the method of Reference [5]. This vector, together with the relative *gross national product* (GNP) of these countries, is given in Table 8.9. The characteristic value was computed to be 7.03, a value close to the rank of the matrix, which points out that the values of Table 8.8 are relatively consistent.

Table 8.9 shows that the values of the characteristic vector are in general agreement with the relative GNP ranks. This illustration demonstrates that the relative weighting method of this chapter can be used to refine a given relative rank value. It should not be considered as a method of correcting completely wrong data, however.

8.5.3 Ranking Sensor Systems

As an example of relative weight computations applied to sensor systems, consider the Table 8.10 relative weight matrix for the five sensor systems given in the numerical example of expert matrix method. The relative rankings given in Table 8.4 were used as a guide in the selection of the relative weights of this matrix.

Table 8.10 Sensor Systems Ranked

Sensor	*TIS*	*MMW*	*IRST*	*TIS and MMW*	*RWR*
TIS	1	2/3	1/3	4/3	2/3
MMW	3/2	1	1/3	4/3	2/3
IRST	3	3	1	4	2
TIS and MMW	3/4	3/4	1/4	1	1/2
RWR	3/2	3/2	1/2	2	1

Using the characteristic vector computer program of [5], we can compute the eigenvector of this matrix as follows:

$$\begin{pmatrix} 1.00 \\ 0.85 \\ 0.31 \\ 1.23 \\ 0.61 \end{pmatrix} \tag{8.16}$$

Note that this vector also represents the relative weight of the sensor systems starting with TIS and ending with RWR. We further observe that these rankings are in line with the previously given values of Table 8.4. The eigenvalue of the 5 × 5 matrix of Table 8.10, which is a measure of consistency of entered matrix elements value, was calculated to be 5.10.

8.6 CHAPTER SUMMARY

In this chapter, methods of multiple sensor effectiveness analysis were described and numerical examples given. Effectiveness analysis methodology began by using analytical, statistical, and computerized methods to develop an expert knowledge base for further analysis. This knowledge base was used in the application of the expert matrix method, where it was quantified into a mathematical procedure. This procedure resulted in ranking the quantitative effectiveness of the examined sensor suites. These ranks were further refined by the relative weights method of T. L. Saaty.

REFERENCES

1. "Westinghouse and MBB Team on Weapons," *Aviation Week and Space Tech.*, January 9, 1984, p. 61.
2. S. S. Blackman, *Multiple Target Tracking with Radar Applications*, Artech House, Norwood, MA, 1986.
3. T. L. Saaty, *The Analytical Hierarchy Process*, McGraw-Hill, New York, 1980.
4. T. L. Saaty, "A Scaling Method for Priorities in Hierarchical Structures," *J. Mathematical Psychology*, Vol. 15, 1977, pp. 234–281.
5. S. A. Hovanessian and L. A. Pipes, *Digital Computer Methods*, McGraw-Hill, New York, 1969, p. 77.

Index